John H. Pinkham

Jack's Memoirs

Visit our Web site: www.ejpenterprises.com.

Pinkham, John H.

2008

Jack's Memoirs: a memoir/John H. Pinkham.

ISBN 978-0-578-00230-9

FOREWARD AND ACKNOWLEDGEMENT

Every Army PX in World War II sold Diary books, so the day I left the country I bought my copy and started writing with a blue ink fountain pen. I was seeing the world for the first time and because military censorship prohibited us from writing home as to where we were or what we were doing, I recorded all the details that are in the first half of this book. The Diary book collected dust over the years and some of my family have never read it. It wasn't until Tom Brokaw wrote "The Greatest Generation" that some of us were encouraged to sit down and tell of those daunting experiences that we considered commonplace and routine.

About that time my daughters Barbara, Patsy and Betty suggested that I rewrite the Diary and include my family history and career involvements. A noted study showed that the average man has at least careers. I was blessed in having a fourth during my business career and into my retirement years. That career was the most fulfilling as it involved the Lord's work and the Lord's ministries. I continued that phase until age 75 and then stepped aside to permit younger men to take up the baton.

I thank Betty Pinkham for solving my computer problems and processing the script to the publisher. I also thank Patsy Drab for her help and guidance. The expertise of Burt Kirkland who produced the front cover for which I am grateful. I also thank Francis Martindale, a writer of missionary books who encouraged me in the early stages to proceed.

Respectfully,

Jack Pinkham

CONTENTS

WAR CLOUDS

When the Armistice was signed in 1918, the "war to end all wars" ended with a temporary peace. In the succeeding years, young people grew up with the uncertainty of their future and of the economy. When the Great Depression came in the early1930s, the stock market had fallen to its all-time low, and the unemployment rate had risen to an all-time high. Those were my growing-up years. As a happy kid, however, I thought I had the best mom, the best dad, the finest house, and "happy days are here again" was the song of the day. So it may interest my family and friends, to hear some of the memories of those days. Some of the family information came from just chatting around the table. Other stories were recollections of significant events. Three of the military years were recalled by pictures, movies, and mementos, but my personal diary was the basis of recounting the year of combat with the 9th Air Force in the Mediterranean theatre.

The Axis was evil after September 1939 when Germany invaded Poland, Japan decimated the Nanking Chinese, and Mussolini's forces moved from their Libyan bases toward Egypt. We then said "thank God for the British" who fought fiercely, but unprepared, against them. Two years later, the British said "thank God for the Americans" when we entered the war. In 1937, with two years completed at New York University, and 19 years old, Eveready hired me as an office boy for three Vice Presidents. In 1940, after the grueling three years of night classes, and two summers, going four nights a week, NYU finally gave me a degree of Bachelor of Science in Economics. This may have inspired a $10 a month raise to $75 a month and a place in the Sales Department of Eveready and Prestone. I thought I had a big job when I got a desk and a telephone.

But the war clouds were forming for the U.S., as Britain and Russia were losing the early war battles. We were all getting draft numbers. My brother Manny went into the Medics of the Army. My British cousin, about my age, frequently sailed to

N.Y, as 1st mate on merchant ships. In his crude Cockney language, he told us stories of the two "bloody" ships that were torpedoed and sunk under him. A few months later, he went down with his ship. The Axis was Evil, so in July of 1941, I enlisted earlier than my draft required and landed in the Field Artillery branch at Ft. Bragg. Providentially, there were two other Christian guys in my barracks. When I told my Lieutenant that I could play Taps, Reveille, and Retreat on my trumpet, he said "get the trumpet" and promised me with no KP or guard duty. When the trumpet arrived from home, he shut off the canned bugle, so I could play for the daily formations. But soon I was sent off to join a regular Army outfit in Watertown, N.Y.

On December 7, 1941, the war declaration sent 100 of us in trucks to guard the power plants of Niagara Falls. I was a truck driver of the big six-wheelers. While there I learned that the Air Corps needed pilots, so I went into Buffalo on my day off, passed all the physicals at the recruiting office, and signed up for a transfer for pilot training. My dad and mom were not exactly thrilled with that news. Neither was the Lieutenant in charge of our Niagara Falls detail. When I told him the Air Corps had priority and I would be leaving, he screamed and yelled, even threatened to court marshal me. But my orders came through for Cadet training.

Maxwell Field was a long way by Pullman car. When we arrived, the first agenda was to get new Air Corps uniforms, white gloves and all. We were marched to a huge dining room, white table cloths, waitresses, and an upper classman on both sides. "Keep your eyes on your plate,"---" "Sit up straight, mister" ----" "No talking" (other than to upper classmen) and it was "Yes, Sir" or "No, Sir" to all upperclassmen. When dinner finished, we were marched back to our rooms, which had a long screen porch, on which we were allowed only a narrow 2 feet. After dinner, about 7 PM, the upper classmen swarmed into our rooms. Made us come to an extreme rigid form of attention. "Stand straight, stomach in, chest out, chin out," commands like that. Then they would pick on us, vocally. "Wipe that smile off your face, mister." And if you broke out laughing, it was dickens to pay. I actually saw one Cadet cry, standing in a corner with an upper classman in his face. One upperclassman, while marching to dinner, ordered me to learn two verses of the Air Corps song. When I couldn't do it in two days he gigged me for five demerits. Enough to be walking Guard on Sunday afternoon, in our Cadet full-dress uniform, white gloves, heavy rifle, and all. They said then it made Officers out of us. The only bright spot was that it would last for only six weeks. Then we would be the upper classmen and could deal it out to a new class. Actually, it was much more fun as an underclassman. During the day, we had studies in navigation, Morse Code, mathematics, meteorology, etc., to get ready for the next air base, where we would learn to fly.

It was another Pullman car overnight ride to our next destination. Mac Peebles, a 6'2" athletic fellow from Tennessee had become close friends. Since my bunk was above his, he took great delight in kicking and bumping my bed from his bunk below, which lasted almost to Douglas, Ga. Wow, what a small place Douglas was. It had been a private field, but now was converted to a Primary Training base, which had the two winged Stearman PT 17's. The schedule allowed for a half day on the flight line, and a half day in classrooms. My flying instructor was a short stocky civilian, Mr. Yadkin, from Brooklyn, who had the power to wash us out at any time. He had four Cadets under his responsibility. He gave us about an hour ride each day, in which he sat in the open-seated rear cockpit, with the student in the front. No radio, just a Gosport rubber tube to speak through. No air speed indicator; just an RPM indicator, plus the oil pressure and cylinder head gauges. Mr. Yadkin could be brutal in that back seat. If he didn't like what you just did, he would yell, and knock your knees with the Aileron stick connected to his cockpit. He taught us to fly by the "seat of your pants". Occasionally, he turned off the engine, and made me do a forced landing on some dried up peanut field. Important to learn to always look for such a safe haven, every time you fly.

So, after 5 dual hours and a lot of acrobatic flying, Yadkin said to me "Do you want to take it up alone"? "Sure" was my shaky reply. That was like a command, because about half of the Cadets never got that chance, and were washed out, sent on their way. So I jumped in the front seat, gave it the throttle, flew around a little to get the feel of it, and came in for a good landing. A good landing in that plane is to avoid a ground loop, as the tail wheel easily spins the plane around, if the brakes are not applied properly. If you ground loop, it is a sure washout. My friend Mac soloed also, and since we roomed together there, we continued to have lots of fun at the Ping Pong table. Mac was a former All-American football guard at Vanderbilt, and very agile. We talked about the Bible occasionally and he was open to learn about salvation and commitment. We enjoyed this little place with its home-cooked, waitress-served, meals. But after 60 hours in the air, some with the instructor, but mostly solo, we had to move on to another more advanced military plane.

So, in July 1942 we were off to Shaw Field at Sumter, SC. Shaw was a big base, large runways, good classrooms, and the fine low-winged BT 13 to fly. All our instructors were Officers. Each of them had seven of us to train in the techniques of flying with a full instrument panel, adjustable propeller pitch, night takeoffs and landing, as well as using the Morse Code in navigation on small cross-country flights. But it was a sad day when Mac Peebles came over to tell me he was leaving the base. He thought he was going to wash out of the pilot program, and didn't think he'd make a good pilot. I wished him well, in the Lord's hands. He had recently told me, he would keep that Gideon New Testament I gave him when he spent a few days in the sick call hospital. It was special because he had signed his name on the place where he made a commitment to trust Christ.

We continued some dual flying with the instructor, but it was mostly solo. We did loops, Immelmans, slow rolls, snap rolls, formations with others and enjoyed racing through the billowy clouds of South Carolina. And there were more Morse Code, navigation, and meteorology classes. Weekends were more relaxing, and I learned that there was a good Brethren Chapel in Sumter for Sunday worship. There I met the John Bramhall family. Mr. Bramhall was a noted conference speaker up North.

I had a few nice dinners at their home, and played some tennis with their 21-year old daughter a couple of times. But it soon came time to leave when we completed 60 hours in that Vultee

Vibrator. And after taking a series of achievement tests, the Air Corp Headquarters told us that about half of us would go on to the next base to be either Fighter Pilots or Bomber Pilots. We did take some achievement tests for that purpose, but when it came down to their decisions, it looked like they just sent the big tall pilots to Bombers, and the small guys to single-engine fighters. Later, highly designed psychology tests did better profiling.

Off in a Pullman car again--- to Spence Field, Moultrie, Ga. There we would be training in the A T 6, a single-engine fighter training plane, with a cockpit full of instruments, and retractable landing gear. An Officer would be our instructor, but once we soloed, he had us flying alone on acrobatics, formations, cross country, night landings, etc. The toughest was a night cross country, on which we flew to the Atlanta Municipal airport, had dinner, and then took off in the dark almost 200 miles to Moultrie. The Navigation system then was atrocious. It used the Morse Code on our low-frequency channels, and the use of Beacon rotating lights for visual guidance. What a relief to touch down in the dark, after that one. We also had to learn to shoot at ground targets, so we had to fly to Pensacola Naval Air Station for that. A 30-caliber gun fired the bullets from the engine Nacelle. The bullets fired in synchronization with the propeller. We all loved the A T 6, as it was nice to fly, and it got us ready for a combat fighter plane after we got our wings.

After we passed all the ground school courses, and flight requirements, we would get our Wings, and be commissioned as 2nd Lieutenants at our final graduation. We had to buy our new Officers uniforms, and pin our new silver wings on it. My mom and dad wanted to come down to Georgia for the graduation, so they rode the train from Penn Station, N.Y., to Atlanta, and the bus to Moultrie. Mom was decked out nicely when they got off the bus, out by our field. I got to put a nice pair of silver wings on her coat. Sadly, that day one of our classmen collided over the field and was killed. And to top it off, I got orders to Cairo, Egypt, to join the 79th Fighter group. So no leave to home and we headed for St.Petersburg, Florida. for combat training in the P-40. C'est la guerre!!!

About 15 of us brand new 2nd Lieutenants arrived at the municipal field in St. Pete., then called Pinellas. The Army Air Corps took it over, put up olive drab tents along side of the clam shell bounded runways, and gave us one brown Army blanket. The damp salt-laden air at night kept those canvas cots pretty cold, so we slept in our lamb-lined high-altitude leather jackets. After studying the cockpit and the flying instructions of the P-40, each of us had to take it off. Solo!! Something you really sweat out. A couple weeks later, I came in on my final approach, behind another P-40 just ready to touch the end of the runway. Though I was gaining on him, I continued to come in hot, rather than give it the throttle, pull up the flaps, and go around. When I landed, the plane ahead, was still on the runway, slowing down, and I was in a pickle. I pulled off the runway, on to the shoulders, which looked like hard packed clam shells-- and raced passed that other P-40 fast enough to get back on the runway to taxi in front of him. When I walked in the Operations room, the telephone was already ringing. " The commanding Officer wants to see Pinkham in his office," the orderly yelled out. So there I was taking a chewing out for what he saw me do, and getting a three week "confined to quarters" discipline. Meaning I would continue daytime flight training , but could not leave my tent every night. Nobody ever checked to see if I complied with the order. But I learned a lesson. Often wondered why-- and what -- was the value of my Gold Lieutenants bars and my Silver Pilots wings those nights. C'est la guerre! With that behind me, Christmas soon came, so I went into St Pete. for a restaurant dinner. After my customary prayer for the meal, a gentleman came to my table asked me if I was a Christian believer, and told me his name was Bill Jelly. I knew his brother Alfred in the Jersey Brethren Churches, and so he invited me to their home for Christmas dinner, and a visit to their local Chapel.

OFF TO EGYPT

When we had flown almost 60 hours, we were supposed to be combat ready. So ready or not, our orders were to get to Egypt, so we headed for West Palm Beach to get all the shots, clothing, and the parachute escape items, for the Libyan desert, where we might be headed. During the three weeks wait for Pan Am tickets, there was a lot of time off. Six of us chartered a deep sea fishing boat at Ft. Pierce and several of us got 7- foot long swordfish. I left mine dead on the dock. My close friend then was Don Cochran, a Clearwater, Florida, resident, already married to Evelyn, a very lovely gal. One Sunday, Don and I attended a fine Baptist Church and heard a pretty young girl sing the solo "The Lord is my Light and My Salvation". It so gripped me that I never forgot it and even sang parts of the tune, later while flying. The microphone buttons on my throat conveyed the sound from the radio to the earmuffs. Frequently, it was strength from the Lord when I could sing those words of Psalm 27.

Since there were only nine of us going to Cairo, we got tickets on a big four-engine Pan Am transport. Don Cochran and I sat together, watching everything. The first landing was at 2 AM in Cuba, because someone forgot to fill the right-wing tank. Then off to British Guyana, then Belen, Brazil, for overnight repairs. Then out in the Atlantic to Ascension Island for more gas. What a desolate rock. No grass, just a runway the length of the whole island. We continued East to the African coast to land at Accra, Ghana. Spent three days waiting for parts for our Pan Am plane. But who cares when you're out there. Nice beaches, nice black men who made our beds, and places to see. From Accra, we headed northeast to Maduguiri, Nigeria. There, more problems, so Don and I walked from the airport, through the tall grass pathways, guided by the throbbing beat of drums and strange sounds. It was a holiday in Maduguiri. Natives lined the dusty streets as racing horses charged to some kind of a finish line. But the atmosphere of evil spirits and liquid spirits, was too much for us, so we walked back through the tall grass, followed by a girl and a boy who offered her to us for "dollas".

Next stop was to be Khartoum, Sudan. Still controlled by the British, there was order, but primitive living, with shops that had nothing to sell but knives and other junk made from discarded steel gasoline drums. But more waiting for parts to fix our Pan Am DC-4. This allowed Don and me to take in a movie at the American Army base. Here, I bumped into Dr. Barnett, whom I had known at Hackensack Baptist. Strange that he was here, as an Army doctor. The year before, he had been on a ship in the Atlantic, with cartons of medical instruments and medicines for the African people, to whom he was going to serve as a medical missionary. His ship, the *Zam Zam*, was torpedoed by a German submarine off the U. S. coast. Most victims were picked up and returned to the U. S. Consequently, this made Dr. Barnett subject to recruitment in the service. Soon he was an Army doctor, and now doing routine duties in dry Khartoum, far from the mission station in Kenya he had intended to go a year previous. We had a great time of reunion out there in Sudan. But the next day, we flew on to Cairo, a thousand miles north, virtually straight down the Nile.

Of course, we could tell Cairo was near when we saw the Pyramids, and the Sphinx, there across the Nile from the big City. All nine of us got together at the airport to decide how to go to the 9th Air Corps Headquarters. But when we debated as to why should we race over there right away, since it took us all of seven days to get there, we all concurred that we should do a little sight seeing at the Pyramids, Cairo, etc., and get back here at the airport in 2 1/2 days, to "sign in" all together. It worked just fine. When we signed in, they sent us by truck to the 79th Fighter Group, two hours away. Landing ground LG 24 is west of Alexandria, on the Mediterranean. Jim Pittard and I were assigned to the 87th, Don Cochran to the 85th and others to the 86th fighter squadron, all of which comprised the 79th Fighter Group, flying P-40s.

The 87th Squadron was still training for combat, by learning tactics from the British Royal Air Force. The airfield had no runways. It was just a square mile of stony sand, outlined by old gasoline barrels. The RAF had been fighting the German Afrika Korps for two years. The British had won a huge battle at

Alamein, just 2 ½ months before, 70 miles from where we were now. Rommel's forces were now on the run in Libya; with Montgomery's 8th Army on their heels. The Afrika Korps had suffered huge losses of tanks and trucks. We would be in position for combat in a week or so, after we got settled. The ground officers, mechanics, and armament people would be leaving tomorrow. The pilots would fly the planes and meet them along the way in to Tripoli.

Libyan Desert 1943.

THE WESTERN DESERT

While a squadron has about 25 pilots, there are frequently less than 25 flyable planes available. So Duke Urich, our Squadron Commander, informed several pilots that we would drive with the convoy to certain rendezvous points on the way to Tripoli, about 1,500 miles away. He said it would be in stages, subject to sand storms. It would be essential for the trucks carrying the ground forces of the Squadron, the mechanics, cooks, radio men, etc., to catch up with the planes, as we moved west. So I was assigned to ride a Jeep, with Captain Bane, the Squadron Intelligence Officer. Basically, there was only one paved highway, as most of the last year's battles were on the desert open spaces, on both sides of the highway. We saw enormous wreckage. Some were British tanks, trucks and planes, but mostly German debris from the recent Alamein defeat of the Rommel Afrika Korps. We saw abandoned fox holes, mine fields, even grave yards, hastily marked with just a steering wheel, a rifle, a helmet, or a propeller to designate what the man's unit was.

We decided to set up the tent and the canvas cot, and it rained until morning. Our progress was slow, as our convoy was crowded by many other convoys carrying supplies to the front. Towns like Benghazi and Tobruk, were by-passed, as they were heavily bombed out and lying in rubble. One good thing, and RAF PX store, was accessible to us and I found a great RAF flying helmet and a dandy sleeping bag. Both served me well. British ships still tried to use some of the bombed-out Mediterranean ports. We passed a ship left burning from the German air bombing the night before. Finally we got to Gazala, where our planes joined us. It took them 2 hours by air, but it took us 3 ½ days on the road. The planes and pilots were waiting, so we could then move westward the next day.

Major Urich told me he had a plane for me to fly the rest of the journey. It was tough to get into the cockpit again. I had not flown for 7 weeks, so my takeoff was jittery, and I didn't have a good feel of the plane. Flying formation was rusty, too. Below us was just the sand-colored terrain of Cyreneca, and soon we came to Marble Arch, a small area marked by a huge Roman arch, since it was then a country border checkpoint to Libya, built in a grandiose style by Mussolini. Italy controlled this part of North Africa, and the German Armies were trying to keep the British from capturing it. Germany wanted all of North Africa, including Egypt, and the Middle East oil fields. They had once been as close as 70 miles from Cairo, at Alamein, and it would have been tragic if Rommel had succeeded.

We gassed up at the Marble Arch field, had lunch at a British mess, waited for a sand storm to pass, and took off west for Darraugh. Visibility was bad; our formation got separated from the other formation, led by Major Grogan, who had the maps. But a very large sand storm had moved in, so we found a field at Missurata. For a desert town, it looked pretty good from the air, so we buzzed it at 50 feet, and landed on the American's part of the field, where they had some B-25's. One of our planes belly-landed. We slept there overnight, took off to our next field where the advanced B party men and supplies would rendezvous. Landed at Gazala in a minor dust storm, but soon the winds and sands were blowing through everything. After dinner the winds increased to 40 mph. with visibility of 10 yards of fine dust. Our faces colored tan, we stayed in our tent to keep it from blowing away.

The next day the storm got worse. Even in the tent the visibility was low. Dust got in our breakfast mess kits. At one time we saw a rainbow of dust. No rain, just dust. Wisely, the crew chief had covered all the planes, and stuffed all the engine openings; exhausts, etc. One of the four pilots in my tent had a radio. About all it received was German propaganda to the British troops. The English they used was crude, like "Damn Americans," to divide the Americans and the British. Cheap try. On Sunday, I went to a Protestant service. The chaplain gave a good message on "friendship with God". Fifty men huddled in a sand-swept tent, while the storm still raged. Finally, it rained, clearing the air, but leaving a lot of mud. From here we would continue training flights.

I kept the same P-40, and started to learn the combat tactics we would soon use on missions: turnabouts, dive bombing, etc. The ground crews were constantly correcting engine problems caused by the sand dust.

From my diary: *There have been some engine failures on take off, rough engines, etc. which make us more vigilant in running our check procedures before take off. Don Cochran, with the 85th Squadron nearby, came over to tell me he was asking for a transfer to bombers. There is a B-24 bomb group at Benghazi, so I wished him well.*

A week later we were still getting muddy looking water from an oasis, 60 miles away. The water truck gave us some in our steel Army helmets, in which we soaped our clothes, and washed our socks. The mess orderlies filtered the cloudiness of the drinking water, so we could make tea. Recently, Dr. Magnus has been giving us lectures each evening on wounds, bleeding and shock for possible emergencies. We have been flying every day, but not long sessions, as the 100- octane gas is scarce out here. We burnt up 4,000 gals yesterday, on several big training sessions, shooting at abandoned tanks, and making passes at a shadow on the desert, made by P-40's flying at 300 feet. Tony Cirrito, our Engineering Officer, bought me a nice scarf in Tripoli for 75 Piastras.

Our next move west was to Tripoli, within operating distance from the front. Castle Benito airport had recently been abandoned by the Luftwaffe, which allowed us to at least to stop there. I went by convoy again; enjoyed seeing the Libyan terrain especially the small town of Gazellten. The green grass and the people seemed like a novelty, after the dry barren desert to the east. There were date palms, olive orchards and some fig trees, due to the proximity to the Mediterranean. Mostly Arabic families lived in small windowless houses. The stone walls were square and the roofs flat. Wells were plentiful here. In the outskirts, herds of starved looking cattle, goats and even camels, were common sights in the pastures.

Occasionally, we stopped where there were children or men with eggs or lemons. We traded them for our sugar, tea, or cigarettes. No money. Solim spoke Arabic, Italian and English.

He offered us dates, peppers, tomatoes and eggs in exchange for our tea and smokes. He showed us his house, just one room of cold brick walls: no beds or tables. They sleep on the floor. We drove on to pass a large fort the Italians had built at Bir Dufan. Next to it was an abandoned airport, which had been used by the Luftwaffe and the Italian air force. The runways had been heavily bombed, the area was still mined, and damaged planes were strewn all around: Stukas, 's, and Machis. Suddenly our convoy had to turn back to Darragh because of lack of space at Castel Benito. A day later we were cleared, and I was assigned a plane.

We landed on Castel Bonito's huge runway and taxied to the dispersing area we were assigned. We soon found our tent, all set up, with a fox hole trench nearby. After dinner the "show" began. Tripoli, 15 miles north of us, suddenly became a maze of bursting bombs, flashes, and searchlights hastily scouring the sky for German bombers making runs on supply ships in the crowded harbor. British anti-aircraft put up a huge barrage, in the semi-darkness. Then there was quiet. The next day we explored the airport, noting that all the hangar roofs were full of holes and abandoned . Shot-up airplanes of the Germans and Italians were strewn around.

Jim Pittard and I took a Jeep into Tripoli. First thing was a haircut and a shampoo. The city was clean, although there were a few buildings damaged. The Italian shopkeepers were nice to us but they had pitiful little to sell. There was a good hotel run by the RAF, where American pilots could rest, so we dropped in for a cup of tea, while looking out on the ship-laden harbor. Many ships were sunk. Around 7 o'clock every night, they are sure to get another bomb raid, as Sicily was an easy distance for the ME-110. Machine guns, anti-aircraft guns of the Brits were spaced evenly along the scenic route of the waterfront. The meager bicycles and horse drawn carriages were crowded by the military traffic. Back at the airfield, we learned that our Group of three squadrons (79th Fighter Group) was placed on special alert for fighting at the front, west of us near Tunisia.

The next day we were told to put on our Officers' Dress uniform, and plan to attend a pilots' meeting over at the RAF

Headquarters. RAF Officers and Commander Broadhurst, in charge of all African Air Operations, welcomed us to their fighting team. He stressed teamwork, radio silence, and the importance of looking around the sky for the enemy. Rommel's Afrika Korps had been decimated, but they were regrouping and replenishing troop tanks and trucks in southern Tunisia. We would be moving up to the front in a week. That afternoon Don Cochran and I went into Tripoli and bought some good date bars, and a second-hand Zeis Camera. We had to use British military currency as liras were worthless. There were lots of soldiers walking the streets. Probably on rest from the front and could be Australians, South Africans, Indians, New Zealanders, or just plain British, as they all comprised Montgomery's Eighth Army.

Back at the field, again around 7 o'clock, we saw tremendous barrages over Tripoli. Red tracers from those harbor guns lit up the early dusk. Then a huge explosion, maybe an ammunition ship, or possibly a direct hit on a bomb loaded German plane. For the first time, wine was brought into our Officers' Club tent. Tripoli wine was a heavy grape table wine. The next day, Don Cochran came over to tell me that he was leaving for Benghazi to join the 276th B-24 Bomber Group. I was saddened to see Don go, but we made a pact that we would get together after the war. Don sure was the finest of men, and lived as a Christian should live. While the crew chiefs were repairing, and upgrading our planes for combat, all the pilots got lectures on intelligence, escape, and capture in enemy hands.

Our tents were not far from our planes which were protected by revetments made of bags of sand. Early one morning at 5 o'clock, we were jarred out of our cots, by a huge crash, just 100 yards away. Then there was some faint hollering from the tragic four-engine RAF Wellington. It had returned from a bombing mission over Sicily and returned all shot up . Fortunately, it hit a sand bag revetment, used to protect our planes, when it couldn't quite make the runway. All the crew walked away from it, probably because the whole tail whipped off, and the revetment absorbed some of the shock.

At 3:30 the next morning we rolled up our stuff, headed for the front, which was now in southern Tunisia. I drove with the convoy again, and found it interesting going through Zivia, Zuara, Ben Guardane, before coming into Tunisia. Arabic people there speak French, and the homes were round-shaped and they were desert poor. We stopped at an oasis village and watched the Arab family irrigate their garden and olive orchard from a well. They used a camel to pull up a huge water filled goat-skin bag. The camel walked down a slope near the well as ropes and pulleys brought the bag up. When the bag reached the top, it fell over a log, as the water ran into a trough 200 feet long through their sandy stony soil. Only one dusty sandy highway was available for our convoy, as we went along the Mediterranean, but we could occasionally see the sea.

الحكومة البريطانية

BRITISH GOVERNMENT

الى كل عربى كريم

السلام عليكم ورحمة الله وبركاته وبعد ، فحامل هذا الكتاب ضابط بالجيش البريطاني وهو
صديق وفى لكافة الشعوب العربية فنرجو أن تعاملوه بالعطف والاكرام . وأن تحافظوا على
حياته من كل طارىء، ونأمل عند الاضطرار أن تقدموا له مايحتـاج اليه من طعام وشراب .
وأن ترشدوه الى أقرب معسكر بريطاني. وسنكافئـكم مالياً بسخاءعلى ماتسدونه اليه من خدمات .
والسلام عليكم ورحمة الله وبركاته ؟   القيادة البريطانية العامة فى الشرق الاوسط

To All Arab Peoples — Greetings and Peace be upon you. The bearer of this letter is an
Officer of the British Government and a friend of all Arabs. Treat him well, guard him from
harm, give him food and drink, help him to return to the nearest British soldiers and you will
be rewarded. Peace and the Mercy of God upon you. *The British High Command in the East.*

SOME POINTS ON CONDUCT WHEN MEETING THE ARABS IN THE DESERT.

Remove footwear on entering their tents. Completely ignore their women. If thirsty drink
the water they offer, but DO NOT fill your waterbottle from their personal supply. Go to their
well and fetch what you want. Never neglect any puddle or other water supply for topping
up your bottle. Use the Halazone included in your Aid Box. Do not expect breakfast if you
sleep the night. Arabs will give you a mid-day or evening meal. Always be courteous.

REMEMBER, NEVER TRY AND HURRY IN THE DESERT, SLOW AND SURE DOES IT.

A few useful words

English	Arabic	English	Arabic
English	Ingleezi	Day	Yome
American	Amerikani	Night	Layl
Friend	Sa-hib, Sa-deck.	Half	Nuss
Water	Moya	Half a day	Nuss il Yome
Food	Akl or Mungarea	Near	Gareeb
Village	Balaad	Far	Baeed
Tired	Ta-eban		

Take me to the English and you will be rewarded. — Hud nee eind el Ingleez wa tahud Mu-ka-fa.
English Flying Officer — Za-bit Ingleezi Tye-yara
How far (how many kilos?) — Kam kilo ?
Enemy — Germani, Taliani, Siziliani

Distance and time: Remember, Slow & Sure does it
The older Arabs cannot read, write or tell the time. They measure distance by the number
of days journey. "Near" may mean 10 minutes or 10 hours. Far probably means over a days
journey. A days journey is probably about 3? miles. The younger Arabs are more accurate.
GOOD LUCK

This "blood chit" we carried in our parachute in case of capture by the local
population.

Washing clothes in my helmet with gasoline in the desert.

FIRST TASTE OF BATTLE

Finally, we arrived at our operations base, just near Zarsis, in southern Tunisia. The field was close to the water's edge, across from an 8-mile wide island. No cement runways, just a hard packed sand like area, which the British carved out for us, along an extended beach In spite of caution, three of our planes nosed over in holes of soft sand, ruining the props. In digging the wheels out, water filled the holes. We already had noticed Luftwaffe planes high overhead, looking us over. Our tents were moved far from the planes, and the first thing we did was to dig a slit trench near them. Each man dug it the length of his body, and deep enough to miss surface shrapnel.(about 2 feet) The local Arabs soon came by to sell us the eggs or chicken they kept under their layers of robes. We bargained using fingers, and motions. They didn't want money, as tea or cigarettes was their demand. In the evening, we could hear the rumble of bombs and artillery, at the front, 20 miles away.

The first mission our pilots flew was to give fighter protection to a British bomb group. Lt. Huff and Lt. Fitzgerald taxied into each other, and chewed up a tail assembly with a propeller. That was two planes out of commission the first day. Every night that week, after dark, we came out of our tents to look at the western sky. The direction from which the artillery flashes came, followed by the heavy bomb noises. Then to the south of us, we saw white burning flares in the skies over British positions that the German bombers were checking out. Most of the noise came from the Mareth Line area, where Rommel's troops formed a defensive position to rebuild their depleted forces. They had over 300,000 men still alive in Africa. On Wednesday, March 11, 1943, Gen. Doolittle and RAF Gen. Broadhurst, with some of their staff, flew in to talk to our squadron pilots. They briefed us on the attack plan and our role in it. Duke later said, joking, "it would reach a crescendo, intensified by bursting bombs and crashing P-40 s".

The next morning we woke up to find that the Mediterranean

seawater had inundated parts of our runways. The British anti-aircraft gun positions and tents had to be moved, but our planes could still use the runways when the wind died down. Jim Pittard and I needed to get more time flying the P-40, so went back to Tripoli airport to ferry some new planes to our squadron. Tripoli was a lot different than several weeks ago. Now the town was busy, women were walking the streets, coming out of hiding after the Nazis troops passed through. Supply ships of the British maritime crowded the harbor. Some sunken and half showing in the water. Got some air time, and returned to our squadron.

Our 87th squadron had been flying some combat missions over enemy territory, and soon I was scheduled to be on my first mission. Twelve of us were to escort and protect 15 British two-engine bombers for a target behind the Nazi lines. I was to be one of four planes surrounding the bombers at their level. Four others were 1,000 feet above us and back, while another four were 1,000 feet above them. Four British Spitfires came up to join us and covered the 20, 000 foot level. As we approached the target, we were jumped by German s, which dived out the sun, onto our guys. The top level engaged them, and Lt. McDonnell was shot down. His parachute opened before he hit the water, and we lost vision of him as we proceeded to the bomb run without any M-109s getting through to the bombers. The Luftwaffe anti-aircraft, poured 88-mm shells at us, exploding with huge black puffs and invisible flying steel. It was on the periphery of the bombers, and it seemed that all of those shells that missed the bombers, came over to us. When we were briefed after the mission, George Lee reported he had bullets holes in his wings, and got a probable shoot down of one of the s. He cried when he found out it was McDonnell who did not return. Lt. Jaslow got credit for shooting down one and Wadi Watkins a probable.

On April 3, Colonel Bates, our 79th Group Commander called Duke Urich, my Squadron Commander and wanted a wing man to fly along side of his plane on a mission. I was picked to protect him on a reconnaissance mission over an area that the Germans were just retreating from, because the British Eighth Army forced Rommel and his Afrika Korps out of the mountains of southern Tunisia. He was assessing the damage. I happened to see Lt. Liggett's plane #84, which was missing in action, lying abandoned, behind enemy lines. There were lots of other Stukas, s, tanks, trucks and artillery debris. Liggett never returned, but our guys were glad to know he belly landed and may be alive somewhere in prison camp.

From my diary: *During this past 10 days I have been on scheduled missions over enemy territory every day. A typical day here is to rise at 6 o'clock, have breakfast in the Pilots tent, and stand ready for a possible takeoff in 30 minutes. One early morning mission, the dawn was breaking, and it was very dark to taxi out to the take off position. Usually, in those conditions, my crew chief jumped up on the right wing, and sat on the leading edge, so he could signal directions and keep me out of the way of the others, as we taxied along. This morning he suddenly started waiving me to stop, which I did. He then proceeded to take off a shoe, while he shook out a scorpion. Quite relieved, he signaled me to proceed to where he could jump off and walk back to his work station. Those critters liked our shoes.*

We had a scare the other day when British intelligence alerted us to intercept a large Luftwaffe bomber group heading our way to hit our air fields. With engines running we sat on the sandy runway waiting for a green flare to give us a "go. When green flare gun went off, we took off four planes at a time, the dust from our props was so great. Sometimes eight at a time, line abreast. When we got in the air, they radioed it was a false alarm, so we stayed in the air, to sweep it, at 12,000 feet, for stray enemy planes, and possibly incite the s to come up after us. But nothing happened. Other days there was more bomber escort missions, involving 18 RAF bombers with our 12 P-40s, at altitudes around 13,000 feet, and the Spitties at 20,000.

Now that Rommel was being forced out of the mountains, he was moving north on the roads to other defendable positions. I had been on an early morning bomber escorting mission, and returned to find my name on a second one which would take off at 4 PM. It was a strafing mission, where we go in low and shoot at every thing we see. The two hours before this mission was full of laughing anxiety. Cheers and prayers for us unlucky ones chosen for this "suicide" mission. We made all kinds of gifts of our personal effects, jackets, knives, watches; anticipating a no return job. Our 12 planes took off with a 500- lb. bomb under the belly, and four 20 pounders on the tips of the wings. Heading over water to the target, we climbed to 8,000 feet, then reaching the land target we screamed down to work over our assigned area. All the time being exposed to the retching anti-aircraft fire of the Nazi guns, I followed my squadron commander down, firing six 50-caliber wing guns all the way. We pulled up, to get out of there, but it was hard to get back in formation as there was so much ack-ack being fired at us. When we landed, we found our whole ground personnel and all the pilots cheering us in and counting us in. Fortunately, only one had to belly land, because an 88mm had shot up his landing gear. Other Squadrons that were on the same effort had much more loss than we did. There were over 100 P-40 fighters on that mission. The 85th reported 5 out of 12 did not return.

Since Rommel was now moving into new positions, in the cover of darkness, it was impossible for the British to give us good targets, so we did some more training flights. One type of training was gunnery, to be more accurate, learn to lead the target, by shooting ahead of it. So that when we pulled the trigger, the enemy aircraft would fly through our 50-caliber shells. We did not have the luxury of a bomber to tow a large screen target, as we had in the states, so we improvised. One P-40 would be used as a "shadow" plane, while the other 12 P-40s would make gun firing passes at the shadow cast by the lone P-40 in the high-noon desert sun. After getting my shots in one day, I was scheduled to be the shadow plane. I flew at the prescribed altitude of 300 feet so the shadow size would be like an enemy size plane. That allowed them to make passes, peeling off from 2,000 feet toward the shadow in the sand, in order to fire all guns at about 900 feet, and pull up for more passes. All seemed to go well, until I landed and taxied over to my crew chiefs. Our usual walk around the plane revealed a long jagged hole in one of the three propeller blades. Our only conclusion was, that a bullet had ricocheted off the sand up into my prop. Prop change needed.

The next day while we were making firing passes over the

shadow in the sand, we came near some Bedouin Arabs with camels: they never looked up. But the final day of all this was marred when one of our pilots did not pull up soon enough after firing his guns at the shadow. The shadow plane had flown over a small lake-sized oasis, which produced a mist that could hardly be distinguished from the water, and Lt. Janus flew right into it. On another day, our squadron spotted an American B-25 which had crash landed near a salt marsh, and some of the crew was out in their underwear waiting to be seen! We reported their location, and they were picked up. We found a wonderful sulphur spring, an artesian well which the native Arabs use for irrigation. I had a great bath in the large tank, big enough for six of us to enjoy the luxuriously warm water.

Pilot briefings for a mission.

ROMMEL'S AFRIKA KORPS

Soon our "rest" period was over, as our whole group was moving up closer to the new combat zone between the British 8th Army and Rommel's Afrika Korps. It was late in the day, when I packed my bedroll and a few clothes for the all night drive to a dirt field 30 miles inland from Sfax, Tunisia. We motored through the Mareth line region, which the Germans and Italians had evacuated. It was intensely cold in an open Jeep. But we got to Sfax at 4 AM, parked, and slept in the seats upright till sunrise. Sfax had taken a horrible beating from our bombers. The waterfront piers, docks, and railroad centers were demolished Three supply stores was about all that was left. I talked to one of the owners, who said he was glad to see the German and Italian troops go, as they stole and plundered food and supplies from those that had so little . Nicely dressed French speaking women and children started to come out of hiding, and the ubiquitous Arabs were still all around.

We pulled into Focconerie airfield, dug a slit trench near our tents, and got ready for operations again. The sandy field was in a nice green area, used by the Luftwaffe and Italian planes just a week before. Thursday, the15[th], eight of us were ordered to fly over the Mediterranean, near Sfax, to protect a British naval convoy of ammunition ships and mine sweepers. As we returned, I noticed Lt. Fred Wright's propeller wind milling, and he soon radioed that he would force land. McArthur and I watched him go in, then came back to report it. Major "Duke" told me to get back in my plane and navigate the search to the downed pilot. Lt. Jaslow accompanied me in his own plane so he could drop a paper message saying a truck was on its way. Fred was okay, and told us later he had met with some fine Arabs who had fed him eggs and offered him lodging. The next day Cardinal Newman flew in to bless us and bless the boys of the ground crews before moving on. Just after supper, while sitting in our officers' tent, two enlisted men came in to tell us there were four German prisoners outside. "Don't worry, they are harmless now" we were told, as we stared and glared at them. Our enlisted men got out some food we never knew existed in our dining tent, gave them cigarettes and treated them well. They claimed they had been without food for 5 days. The one who spoke English claimed Hitler would still win the war. Following the meal, they were turned over to the British for interrogation.

As the 8th Army advanced, we did too, so again it was pack up and go, and we moved to another abandoned air field near Kairouan. From the air we could easily spot the coliseum ruins around El D'Jem, left from the days of the Romans and Carthaginians. Kariouan was an ancient center of Mohammed. Three pilgrimages to Kairouan redeemed the same heaven as one trip to Mecca. All the streets were cobblestones, heavily trafficked by cloth-face-covered women and robe-dressed Arab men. The airfield had been hastily abandoned only a week before, and there were some damaged Messerschmitt and Focke Wolfe 190's left behind. This field put us closer to the top part of Tunisia where two large cities, Tunis and Bizerta, were the chief supply ports of the German and Italian armies. They were rapidly losing ground. The American forces which had come from the Casablanca landings, were now engaging in combat and helping the Montgomery 8th Army squeeze the Axis to the sea. But they were green and we heard stories of the American inexperience and snafus. We were glad to be working with Montgomery's forces, as they really knew what they were doing, since they had already been two years fighting there in North Africa.

So operations started immediately. Now it was fighter sweeps up to the Bay of Tunis. British Air Intelligence had secret information the Luftwaffe would be sending swarms of heavy cargo planes across the Mediterranean from Sicily to supply their Afrika Korps. Our job would be to intercept them on the way in to their Tunis airfields. We patrolled the area, sharply looking around the sky, until our fuel got so low it forced us to return to base, 100 miles away. As we left the area, other squadrons replaced us. Sure enough, the 57th P-40 fighter group, which replaced us, spotted those huge engine German transports down below them, protected by dozens of s. The 57th poured down into them as the enemy tried to disperse and dive towards the trees.

With the help of other squadrons, 45 Ju- 52s, 7 s, and one ME-110s were shot down, one of the heaviest air victories in the war. We had just missed it.

I was on similar missions in succeeding days, but no more sightings until the day South African P-40s ran into another bevy of Ju-52s and shot down another 10 planes. On another day they also shot down 31 big ones, 7 ME-109s and an Italian Macchi. The day after the 57th's big victory, German night bombers flew over their field, dropping bombs two nights in a row, damaging planes and tents. This convinced us that German Intelligence knew all about the 57th and us. Since their field was only 5 miles away, all of us made sure we had slit trenches near our tents cleaned out and ready to jump in.

It was obvious the Axis armies could not get supplies across the Mediterranean by boat, as the British Navy now controlled those waters. I flew with the Squadron on another dive-bombing run on a target deep in to Cape Bon. As we cruised up the coastline, I noticed my shoes were all wet with oil and an inch of oil was on the cockpit floor. In a fighter plane, your feet control the rudder by cables, as well as the brakes. The loose oil had to be coming through the engine wall. I had to get down before it caught fire, so I radioed my formation, dropped the bomb in the sea and headed home alone. On my approach I cut the switch, made a dead-stick landing, and just sat out on the runway until the trucks brought me in. Later the crew chiefs found the oil reserve tank cap had not been screwed on. When our pilots returned from that mission, they reported shooting down 5 ME-109s, after I had to pull away to fly back.

Sunday, April 25, was my busiest day so far. It started at 6 AM in the dark. Went to briefing at the operations tent, started up the engine, and exactly at 7:05 AM, 12 of us took off to rendezvous with 18 bombers. We flew protection for them at 13,000 feet, experiencing the worst, most accurate ack-ack we had ever felt, from the German 88 mm gunners. All we can do is to weave, dive a little, and climb, to throw them off somewhat. None were shot down that day, but two had been, the day before. That afternoon I was on another bomber escort mission to the Tunis area. On this mission, 50 of us were escorted 36 bombers.

Again, no losses, but there was intensive 88-mm shellfire at all of us. We landed before dark, had a meal of more Spam, and went to the Chaplain's tent for the Easter service, and communion, using the Methodist prayer book. After that there was a meeting in our Officers' tent, for a briefing on the last phase of the battle plans for the final defeat of the enemy in North Africa. Actually, it started that night, as the Tank, Infantry and Artillery troops start to encircle the Germans and Italians. The plan was to avoid damaging the cities of Tunis and Bizerta.

In some free time the next day, it was my turn to take a flight in the fixed up German fighter, the ME-109. Lt. Jess Jory and some crew chiefs had found the plane after it had belly-landed for some unknown reason. Jess could read German, so it became his project. From another abandoned plane, they got a good propeller and two wing flaps, and the plane became our squadron prize. Wright Patterson Field in the States wanted it sent back for testing. But our pilots said, "after us, you come first" and we all started test flying it ourselves. Since there were no instruction manuals, the last one to fly it, passed on the procedures to the next pilot. And as all the instruments were in metric, the critical airspeed indicator was therefore in kilometers. I studied the cockpit, learned the take-off intricacies, and barreled down the runway at 90 mph until it lifted off. There was a landing gear lever I was supposed to move to bring up the wheels, which I found unmovable. I concluded that the landing gear was stuck down and merrily continue my flight, staying close to our field so I would not get shot down by someone who spotted the Swastika. In my approach to land, since there was no tower and no radio contact, I steamed in with plenty of extra speed in kilometers. When I hit the ground, dust and dirt flew all around as the prop chewed it up, and I was obviously sliding on the belly to screeching halt. Thank the Lord, it did not tumble. Our guys were laughing and sympathetic, but I was apologetic. The nicest thing my squadron commander could say to me was "that's OK, Pink, now we can telegram Wright Field and tell them the ME-109 is inoperative." Actually, within a few days, the crew chief mechanics found another prop and a set of flaps to have it flying again. We loaned it to the 85th Squadron pilots and they all flew it too. And now that we knew the weaknesses and strengths of the Luftwaffe's best plane, we changed our flying tactics, should we encounter them. Never try to dive away, or climb away, but stay and turn in circles to get on their tails. But I'm not sure if that ME-109 ever got to Wright Field.

At 3:30 that day, we took off with 500- lb. bombs under the belly, rendezvoused with four Spitfires of the RAF and headed for the water near Cape Bon. At 13,000 ft. we spotted two ships. One was a white German Hospital ship, which we were not to touch, and the other was German supply ship, which was our real target, on which to release our bombs. On the way back home, I saw a plane heading directly in front, and spraying bullets all above my canopy. I could see the red fire from his tracer bullets, which are one in every four 60-caliber shells. I knew it was one of our planes, so I sharply pulled away to the right, and watched for his number as he flew by . "Hey, what are you doing to me", I yelled over our VHS radio frequency. When we landed, I accepted a profuse apology from Erwin , one of our new pilots. C'est la guerre, and thank the Lord!! Not a day goes by that I don't pray for safety.

In the evenings, pilots and administrative Officers, which we call ground Officers, get together in a large Club tent. Lately, Dr.Magnus, Duke Urich, my Squadron Commander, and Col. Grogan had been coming over to my tent for a good cup of tea. I had a good Primus stove, which can use the same 100-octane gas used in our planes. It looks like a tea kettle with a pump to pressurize the gasoline, Doc Magnus is our Squadron Flight Surgeon and treated all our minor ailments, and watched the pilots for signs of fatigue. He had been a gynecologist in private practice, so we learned a lot about female anatomy!! Military rank among pilots in the Squadron meant nothing on the ground, when we mingled as friends. Our tea sessions sometimes drifted to Bible discussions, and meaningful thoughts about our Creator and Redeemer. My Bible was a great source of strength.

The last 10 days had been very busy, sometimes flying two missions a day over enemy lines. Duke asked me to take a ride on a DC-3 to Tripoli, Libya, to pick up some repaired P-40s. So four of us spent the night at the RAF Club in Tripoli, nice bed and hot bathtubs. Returning the next day, my engine cut out going down the runway, so I turned back for another plane. While waiting to arrange all of this, a severe sand storm blew in, with a temperature of 105 degrees. Flew back alone and glad to be back in the cooler elevations of Tunisia. Meanwhile, Montgomery was advancing . His plans were to push through the valley region towards Tunis, while the Americans held the Germans off, west of Biretta. Lately, our missions had been directed toward German ships and docks. We sank a destroyer in the Mediterranean, and got an assist on another one. Enemy fighters were not coming up after us now. They seemed to be using all the planes to attack the ground forces, so they came out to drop bombs and then run.

But we were going deeper into enemy territory, as on Sunday, Mother's Day. We got up at 6:00 for a 7:15 rendezvous with 18 bombers heading for the island of Pantelleria, off Sicily. We expected to see lots of enemy fighters so deep in their territory, but we only got anti-aircraft shots bursting all around us. The heaviest yet experienced. The bombers we escorted were very accurate and did a real job on that very large airfield. More dive bombing missions on trucks, trains, and supply ships, some with 500 pounders, plus two 40-lb. bombs on the wing tips.

One day my electric system failed over enemy territory, and I had to leave the Squadron without radio or prop pitch control, 75 miles out. Propeller revolutions were excessive for landing so I cut the switches and made another dead-stick landing. The 57th Fighter Group, a sister group of my 79th Group, had spotted a destroyer at sea the week before. They made some dive bomb runs on it and got a direct hit plus some near misses, only to find it was a British destroyer. Fortunately, the bomb never exploded, but it slid down the side of the stern. The 57th entertained the whole British destroyer crew at the 57th's field a few days later.

Rommel, Montgomery, Eddie Rickenbacker

P-40

300,000 ENEMY CAPTURED

Finally, the capitulation of the entire German and Italian Armies of Africa was announced. Scotty, one of our pilots, was out in a Jeep near Enfidaville, when the crack German 90th Brigade surrendered. Heavy fighting had gone on till the last, until British bombers came over the hills with some bombs. It wasn't until then that the Jerries and the Eyeties came running out with white flags. Our squadron stayed on alert for take-off should the enemy try to escape by air or sea back to Sicily, but there were no calls for help. Some 300,000 officers and men were captured and started walking in all along the front. All their guns and equipment were abandoned as they straggled the roads on foot, waiting to be sealed behind barbed wires and be fed. We didn't have go far to see them and talked to those who spoke English.

The next day a DC-3 landed at our field, and taxied over to our end. In the party was Eddie Rickenbacker, of World War I fame, who simply came to cheer us on and sign some autographs. He told us how tough it was for the boys out in the Pacific against the Japs, and a bit about his rescue on a raft after over a month in the water, following engine failure over a remote part of the Pacific. The amazing part of his visit, though, was how he found our field. In the arriving party was Lt. Hamilton, one of our pilots who had been shot down just over enemy lines by ack-ack.. After his engine was hit, he headed toward our friendly area and belly landed in an open field about 10 miles inside our lines. He stayed near the plane, until today, when he spotted the DC-3 flying overhead. He radioed to the DC-3 pilot , who, with Rickenbacker, agreed to try and land near him in the open field. After landing, the plot asked him if he happened to know where the 79th Group was located, to which he responded "Sure, that's my outfit". He then navigated them to our landing ground. Rickenbacker told this to us, pointing upward to the "power above" in situations like this. I got his autograph on my "short snorter "paper money.

With no missions scheduled, Jim Pittard, Owen, Hennin and I

took off in our individual planes for a day's visit to Constantine. Our 310-degree heading didn't seem to reach there, so we radioed each other, and agreed to land at a small strip that had some planes on it, and get a new heading for Constantine, Algeria. The area was beautiful, with mountains and gorges all around. Constantine had an attractive arch bridge over a gorge in the middle of the city. The American officers at the field loaned us a command car to drive into the city. We took off for home at 5PM and got back at 7 PM. The next day, three of us took a Jeep on a tour of the captured area on the way to Tunis. We encountered detours, mined areas, and general destruction. Prisoners were walking on the roads from which they had abandoned their trucks, tanks artillery, some burned out, some still usable. Guns, ammunition, tools left strewn around. Came to Hammamet, Bon Ficha, and then to the outskirts of Tunis. Tunis was now crowded with soldiers of the British and American fighting units. Nearby was the town of Carthage, where there were picturesque scenes of the ancient Roman Aqueduct which carried water 30 miles from the mountains in the visible south region. On another day I took a Jeep to look for an English cousin who supposedly was with an outfit around Sousse. To no avail. So we explored the ancient catacombs there. Built by the French, these tombs still contained bones that lay out in the open. Some tombs were underground passages miles long.

Then Duke gave three of us a 3-day pass, so we hitched a ride on a DC-3 that was going to Algiers. We had orders to be on leave, but no reservations. All the billeting quarters and hotels were crowded with soldiers who had assignments. Luckily we found a GI who let us come to an apartment that would be available for just one night. The other two nights we slept in Pullman car bunks at the edge of town. Algiers a French speaking city, was a miniature Paris, undamaged, but overrun by war. Jim Pittard, Bud Hennin and I walked the streets, had some good French meals and all the ice cream we could eat at the Red Cross club. While looking for rooms, we met a woman on the street who said in broken English, part French "Yes ,we have rooms". So we followed her up some stairs to the 2nd floor and the "rooms," only to find out quickly that it was a brothel!! Sunday morning I went to GI Post chapel service. Later, saw

Gen. Eisenhower and some officers drive by in a staff car driven by a good-looking WAC. Also saw Tom Harmon, a former All-American Quarterback. Met some pilots I knew and even bumped into a friend of John Goldie from Yonkers Chapel. Getting out of Algiers was worse than getting in, and we were now 2 days past our time. At the airfield, there was no DC-3s that would take us with our # 3 priority, because the colonels and majors had # 2 priorities. After talking to several pilots we found one who could take us to Constantine, where we got a ride on a B-25 to Sfax. We slept at the 12th Bomb Group headquarters overnight and found a DC-3 going in the direction of our Kairoun field.

On arrival three days late, there were no planes around and half the outfit had moved 100 miles north. We took a Jeep to join the pilots, at our new landing ground. The news now was that we would continue operations under British orders, still operating under the 9th Air Force HQ in Cairo, and we were pleased about that. Our missions now took us out in the Mediterranean, where some strategic islands (Italian) had been the Germans air supply link to their African units. Pantelleria had had a steady bombardment, in preparation for a land invasion. I suffered a tremendous sinus head cold and missed action for a week. But during that time, our pilots were protecting some Bombers, when they spotted Spitfires tangling with some ME-109s. Five parachutes were seen floating down. Then Hennin spotted a large German fighter group near the water. That afternoon, 15 enemy fighters were shot down. With my earache, all I could do was watch them land and hear their stories. Goose Gossick got one, Adair got one, as did Barrinati, and Kirch. It must have been terrific air combat. But where is McArthur? And Anderson? He was seen to get three ME-109s and then stayed, circling over Mac, who was in his rubber raft in the water. Goose and Jess Jory gassed up their planes and flew back to where Mac had parachuted. Meanwhile an RAF amphibious Duck was also heading to pick him up, while Goose and Jory gave protection. The Duck brought Mac to a Destroyer for overnight. Mac returned the next day and we were all eager to hear his story. He shot down four ME-109s and one Italian Macchi. The Jerries had lost 25 to the 79th pilots in the last few days and no losses to us.

The next day, June 11th, Pantelleria capitulated. After more bombardment in the morning, a British landing force hit the shores. The Italians remaining there, put out the white flags immediately, and gave up. The island had taken a beating. Our job was done.

Our three squadrons of the 79th were ordered to go off operations and move again, back to Zarsis, in southern Tunisia. Several hundred enlisted men and ground officers drove off in trucks. Duke, now a major, told the pilots to take our planes anywhere for a couple days, while the trucks were on the roads to Zarsis. A few of us went to Tunis, stayed in a nice hotel, now run by the 12th Air Force. Rooms and meals were free. And Tunis turned out to be an interesting French city, but not as clean or classy as Algiers or Alexandria. To get "home" we flew our planes down to the desert area , where we started our missions, and where it was really hot. The word got out that the next big push would be the invasion of Sicily or directly to Italy. In the meantime we would get new planes, rest and even get a few days leave in Egypt.

We were excited when a DC-3 came by our field, picked up 34 of us, and flew us 1,500 miles back to Alexandria, Egypt. The Red Cross picked us up, and brought us to their hotel on the seacoast. My room faced the water and the roar of the waves. Our dinner was at an RAF officers' club, a great change from Spam and bully beef. The next day, I went to a nearby Church Sunday School, where a lady told me of her friend who fellowshipped at a Brethren Church, but I would have to go to the Scotch Presbyterian Church and meet her husband, an Elder there, Mr. Givan, and get the directions. I took a taxi, enjoyed the morning service, and met Mr. Givan. He said "come to our home for tea at 4:00", so later my taxi pulled up to a long entrance to their mansion. Mr. Givan was in charge of Lloyds of London for Egypt. After tea we went to the evening service for the Lord's Supper with the Brethren. Afterwards, an Egyptian family invited us to their home for a light supper. Professor Feccuri taught chemistry at the Alexandria University. Pure Egyptians have their own special pride, and don't want to be called Arabs, which they are not.

The next day a few of us took a train to Cairo. Did some sightseeing around the Sphinx, Pyramids and the Bazaar. We also enjoyed those good hotel beds, and good meals. The Sphinx still had the sand bags all the way from the ground up to the chin, for support if bombs were to drop near it. The Egyptians had been expecting Rommel's Afrika Korps to break through, prior to the battle of Alamein. Back in Alexandria we took some special city tours, to see Ptolemy's pillars and excavations, where digging for Alexander's bones was going on vigorously. The most interesting was Egypt's summer Presidential Palace, in which we saw the dining room table, and silver enough for a sitting of 50. The bedroom walls were covered with silk damask and the beds with fine silk over down feathers.

The nine-hour flight back to our base, took us by Benghazi for gas. I tried to call Don Cochran, but could not reach him. Some pilots had to go all the way to Dakar to get new planes that had arrived from the States. Temperature there was 105 degrees, some days 112. The Arabs brought us watermelons, and we went often to the beach to cool off. At the field, Joe Specht one of our new pilots, took off down wind on a the short part of the field, but couldn't lift off in time, totaling the plane, and himself, as the tail was wrapped around the cockpit on the impact.

On July 8, Duke asked me to ferry one of our older planes to Cape Bon, and pick up a new model P-40L. What a scary trip. While out of sight of over a large bay, my engine started cutting out. I cut the throttle setting and the prop pitch back so it ran smoother, but the sputtering continued. I decided to navigate to the closest shore line and look for an airfield, anything. Fortunately, I soon found a US bomb group field, landed, and asked the crew chief to try changing my sparkplugs and install new magnetos. The engine checked out fine on the ground, but as soon as I headed north again, it sputtered and kicked again. I decided to just cut back the settings and tough it out for one more hour, and chalk it up to those nasty sand storms. That night I stayed with the 325th Fighter Group(also P-40s). I got in on a special meeting that informed them of the invasion plans that would be in a few days. Over 2,000 Naval and supply ships, submarines, glider troops, paratroops, all were to land on the east and southern coasts of Sicily. The 325th and the 33rd Fighter Groups would be supporting the American troop landings. The 79th and the 57th Fighter Groups would be supporting the British 8th Army on the southern corner. So the next day I flew back with a dandy new model of the P-40, which Duke said would be my personal plane. My crew chiefs put my name on the fuselage, and a painting of the squadron insignia (mosquito with a machine gun) and my ID "the Flyin Trapeze". Our stay in North Africa had been restful, lying around in shorts, shirtless, swimming, lounging on our cots in our tents out of the hot sun. Shortly after we had settled here, a chicken laid three eggs under my cot, and freely came and went to sit on them night and day, as our tent was open all around the bottom. My tent mate and I counted the days to hatching time, but now that we were pulling up stakes, we wouldn't see those chicks hatch. That day, Charlie Loundes spun in, doing a tight Luftberry turn, right over our sandy landing ground. We heard the crash, ran over and helplessly watched as he and his shattered plane burned away. The Chaplain conducted a short service, later in the day.

Captured German soldiers and captured airplane.

Captured German VW jeep.

Sphinx chin supported with sand bags in anticipation of bombings.

INVASION OF SICILY

Lots of activity getting ready to leave again for a new field the British would prepare for us the day after the invasion troops land and secure the area. Our trucks and ground crews would land in Sicily by boats. Our planes would land at Malta, and be ready to help in the invasion. Our guys were some of the best pilots around, but some times it didn't mean a thing on a strafing missions, or dive bombing mission, when the ack-ack was heavy. Now that the Luftwaffe had fewer planes down there to come up after us, we were getting more of the dangerous, close to the ground missions. Colonel Grogan came back to tell us the invasion had started and so far had been successful, except that the American anti- aircraft had shot down 23 American DC-3's loaded with paratroopers. So sad.

The next day our entire squadron took off to a cleared airfield, on the south-east coast of Sicily. Hundreds of supply ships, at anchor, crowded the harbors the Germans and Italians had evacuated. Our arrival that night was greeted by a one-hour blitz by Luftwaffe bombers, targeting the ships in the nearby harbor. Also scaring us to death, and driving us to our fox holes. At 5: 30 AM, we took off with 500-pound bombs for a dive bomb mission. We rendezvoused with the 85th, so we had 24 P-40's heading for the Mt. Etna and Messina Harbor region where the enemy was bringing in supplies, and digging in for a strong stand in Northern Sicily. After returning, Mac and I took a look around the area in a Jeep. While we were gone two s strafed our field, and scared our whole unit to death. Two planes were damaged, one of which was mine. New engine needed. The rest of the day everybody was jumpy, and we went to bed knowing that jerries would be coming back tonight. And they did, about 11:30 PM, a terrifying raid again at the ships in the harbor. We were only a few miles away, so the brilliant, phosphorus parachute flares they used to light up their targets, drifted over above us, by the wind. We jumped in our fox holes, the bright flares illuminating our souls.   Everyone just laid there until the British anti-aircraft batteries protecting our field, stopped firing, and the jerry bombers droned away.

The following morning, Maj. Uric (Duke) came walking down to my tent a 6:00 AM. "Lydia," he called me, "Want to go on a mission in 15 minutes"? Of course I would go, although I had not seen my name on the board the night before. I was usually his wing man on all the previous missions and he always was the lead plane and navigator for the squadron, as he was on this one. Duke led us north along the coast to the Straights of Messina and Messina Harbor. There was plenty of shipping targets in the two-mile distance between Messina and the toe of Italy. At 13,000 feet the enemy 88-mm shells were breaking all around us. Big black puffs meant heavy flying steel. Shrapnel ripped holes on my plane's left side and blew my hand off the throttle position. It is rare to hear them explode near you, but that time I did. Duke peeled off, the first to go into a near vertical dive, while the rest of us followed him down. That time I partially blacked out on my pull out after bomb release, and went screaming out over the water, to join up with Duke, and some others, for the trip back home.

The 99th Fighter Squadron arrived at our field, the first Negro Fighter Squadron of the U.S. Army Air Corps. Later, they became known as the Tuskegee Pilots as they were all trained in the all-black flying program at Tuskegee, Alabama. They were camped at the other end of our field, had their own P-40s, mechanics and complete organization. They would be flying missions with us. We got to meet some of their pilots, and soon learned these men were the cream of the crop. Not only were they good pilots, but they were hand picked men for future leadership in other black flying groups. The plan was for them to stay with us, learn our tactics, go on the same missions, and then return to the U.S. to form new black fighter groups. We were not encouraged to fraternize with them, however. But we did cooperate well in radio communicating, in rendezvousing, etc. en route to a target.

One early morning mission, my 87th Squadron taxied out for takeoff, and we sat at our end of the runway, engines running, and waiting for the 99th Squadron to take off before we did. Since there was only one runway, they took off from the other end of the runway, toward us, and most of them got airborne in pairs. But the last pair came drifting off towards our planes. And they were not lifting off as the others, but were heading for my plane and my wing man's plane. Finally, they lifted off, barely enough to fly right over my canopy, but not enough to miss my wing man's propeller. I saw his prop fly off to the ground, as the black pilots landing wheel tipped it. Luckily he continued to fly but he had to be checked for damage before he was allowed to come back for an emergency landing. That was one of my closest calls. Within a month they all packed up and went back to the States, and formed three more squadrons known as the 99th Fighter Group, and contributed substantially in Europe.

We have been kept busy, and there have been some direct hits on supply boats, even a destroyer was sunk recently, but we also see a lot of splashes in the sea... Our P-40 Fighter groups continued these dive- bomb runs on Messina Harbor for several weeks, because it was just too risky for the heavy and slower bombers. Command stopped sending them there. Our planes have been shot down by the ack-ack too, but we are more expendable, since we have only one pilot in a plane, compared to a crew up to nine men in a bomber. My best friend in the squadron was shot down and never returned. Jim Pittard's engine was hit. He was seen to bail out over water, open his parachute, and when in the water, inflated his little rubber boat. He waited for a German rescue plane to pick him up (six months later he sent me a post card from Stalagluft #4 near Berlin). Liggett and Highfield also became prisoners of war. *I keep thanking the Lord every day, and still find great peace in Psalm 27.*

We needed some more planes to replace lost ones, so Detrich and I were flown back to Africa to pick them up. On the way, our DC-3 stopped at Malta, filled up with gas, had a cup of tea, and took off for Tripoli. After a night at Hotel Gedink, we got a ride in Gen. Stricklands' DC-3 to Zarsis, down in southern Tunisia.. I checked over the new plane and flew straight back to Sicily. It is now Aug 7th and it appeared that Montgomery's 8th

Army and the American Forces were gathering for a big push to drive the Germans and Italians out of Catania and across the Messina waters to Calabria on the toe of Italy. We had a big mission that week to Catania which was an all-out strafing of tanks, trucks and bridges. We went in four plane raids. Anderson, Owen, Adair, and I zoomed down the side of Mt. Etna to the roads around Catania, firing our six 50- caliber guns. Then Adair radioed that his engine coolant had been hit by enemy shells. Andy pulled out and guided Adair back toward friendly territory, where he bailed out. I continued hitting trucks, etc. and headed for home alone. Later, Owen came back with his engine and wings all shot up, but he made a miraculous forced landing. Another one limped back the same way and the pilot bailed out right over our field... The next day, there were more attacks on the Catania railroad and signs the Germans were moving out. Not surprising, as the Allied infantry had been pressing on the rear guards, to force them over to the toe of Italy. The Luftwaffe was not very visible in the daytime, but they were out at night. That week, while just getting off to sleep, I saw from my open tent, a barrage of anti-aircraft fire a few miles away. It was ME-110s and Focke Wolves dropping bombs on a RAF airfield near Lentini. Ten men were killed and forty wounded, plus twenty of their aircraft were damaged heavily. Fortunately, they left us alone.

Catania was now won. The next day four of us took a truck up to Catania to look over the scene. Troops were still taking their positions, roads were clogged with tanks and men on the move. Water mains leaking, some buildings still smoldering and streets littered with debris. The airport had been heavily bombed. The people were just returning, and some were exuberant we were there. They booed at pictures of Mussolini or Hitler. The next week was hectic: strafing missions every day. The Luftwaffe was seldom coming up after us, but occasionally we caught them on bombing missions, and they took off and ran. But we continued on strafing runs, which we still feared. I was on two one day. At 5AM we got out of our sleeping rolls, had some coffee, went to the pilots briefing by the big target map, and heard Capt. Bane, our Intelligence Officer, outline the objectives, and the navigation headings. We printed those headings on the back of our hands, in black ink, so we could see them when under stress. Then put on our parachutes, which included the heavy rubber boat , and rode to our planes in a jeep. Mac led our squadron of 12 planes to the targets beyond Catania and closer to Messina Bay , where the enemy was trying to escape in the dark nights, to the toe of Italy. We headed for Mt. Etna, flew right through the fumes of the smoldering crater, and slid down to the winding roads between Franklinville and Novarro, firing hundreds of rounds at trucks and tanks on the move. They put up so much of their 40-mm shells, we could only make one pass and get out of there.

Returning with George Lee on my wing, we landed in pairs, to cut down on the dry dust of our dirt field. Lee forced me over to the right after we landed, where my wheel collapsed by a pile of rocks, spinning the plane around enough to clobber the wing. So my nice running plane had to be sent off to the "repo-depot" for a new wing and a landing gear. A new P-40, #82 has been assigned to me. The next day, four of our planes were jumped by 12 ME-109s. Macarthur, Watkins and Scotty got in some good hits, but Bud Henin radioed that he had to bail out. Anderson followed him to the water, despite a bleeding left arm he received from enemy bullets. Anderson called the RAF for help, to rescue Henin. Two Spitfires came over, protecting the amphibious duck. One Spitfire was shot down by shore batteries, but the duck found Henin, and by miraculously dodging heavy flak, flew out of there. The duck had engine trouble after getting in friendly waters, dropped Hennin off at a gunship which dropped him near a hospital. A couple days later Hennin joined us, with scores of wounds to heal.

Here in Sicily, there were some classy villas owned by wealthy grape orchard owners. Our pilots negotiated with the caretaker, to allow a temporary occupancy for us, as we passed through. The estate consisted of several hundred acres of grapevines, orange trees, date palms, and fig trees. One of the curses of southern Italy and Sicily was the absentee ownership of the grape orchards.

MALTA REST   ITALY SIEGE

The Axis armies were moving rapidly by night, deeper into Italy, and the temporary lack of targets had slowed down our operations. Duke told a few of us to get lost, so Hennin, Macarthur, Dzamba and I took off for Malta in a DC-3, for a six-day rest. We found a nice hotel in Valetta, ate some nice meals in a friendly atmosphere with families, British Naval officers, and even the Chaplain General of Malta, Col. Gethin Jones. The first day we walked the Valetta streets, I bought some beautiful Maltese lace for my mom . The lace was not made with linen, as there was none, but with parachute silk from the scores of parachutes of bailed out pilots, during the two year siege of Malta by German bombers. We toured the inside of St. John's Cathedral, which was huge and beautiful. The walls and ceiling were covered with paintings of renowned artists. All the sculpture was covered up with sand bags. The side Chapel room had been hit by a bomb. The past two years German Stukas dive bombed Valetta harbor day and night, because so many British Naval ships used it as their main base from which to control the Mediterranean. Consequently, Malta became the most bombed place in the war, defended earlier by obsolete planes of the British, against the very effective modern German Luftwaffe. Sunday, we took a guided tour to the little island said to be where Paul, in the first century, was shipwrecked.

Later I visited with an English missionary, John Baker, who also worked at Barclays Bank .He took me to an Officers' Bible study that afternoon, and the to a Brethren Chapel , where they had the Lord's Supper at 6 PM. Baker took English Capt. Armstrong and me to dinner. It was interesting as we were joined by the brigadier general, who commanded of all anti aircraft batteries on Malta. A fine Christian. The next morning, the British aircraft carrier *Illustrious* was in the harbor, so we went over to see all the Seafires, Barracudas, and Wildcats out on the decks. The pilots showed us around and treated us royally, and for our arrival aboard, the crew stood at attention, and bugles blew. We didn't know how to handle this stir except to salute the

Quarterdeck, where they were. Later, by appointment, we visited the battleship *Nelson*, looked at their amazing 16-inch guns, as well as the 8 inch and 40 mm. The following two days, the *Nelson* and the *Illustrious* steamed out of Grand Harbor, joined up with other cruisers, destroyers and carriers for some big sea engagement, probably, the coming invasion of Italy.

John Baker picked me up to go for a swim at the beach. Malta's beach had no sand, just hard, fairly smooth lava. At dinner in the hotel some others from our squadron were there, and told us that Duke wanted us back. So we arranged for our return flight. That night the Maltese sirens screamed in the darkness when the lights went out and people scurried to air shelters. We went up on the roof to see the "show". Shortly, up at 12,000 feet, there was a huge flash, a plane had been hit and was a flaming torch diving into the sea, exploding before it hit the water. After the sirens gave an all clear, we learned that a German Ju-88 bomber had been shot down by a Mosquito, a great night fighter the RAF uses. At the airport in the morning, a Mosquito crashed on landing. When our plane arrived, the pilot brought us the big news that the Italian Air Force was giving up. Radio news confirmed that Italy had surrendered.

When we got back to Sicily we were rushed to a pilots' meeting to instruct us not to shoot at any Italian planes in the future. I flew three missions in one day, and two each day as soon as I came back. It as obvious there was an evacuation of Sicily by the Enemy troops. A DFC recommendation was sent into Cairo for me, but I would believe it when I saw it. We continued to operate from Sicily until an airfield could be prepared for us up in the toe area of Italy. One of the last missions from there, George Lee and I were diving into road targets, when my guns suddenly jammed. Seconds later, George called me, saying his engine coolant had been hit. I stayed with him, guided him south toward home. However, he abruptly turned north, across the water to the friendly part of the toe of Italy, where he belly-landed safely. My 51st mission.

We finally got orders to move to an airfield in Italy. Capt. Bane made arrangements for the officers to stay at a pretty white villa. It had 35 rooms and a large courtyard with white marble statutes all around. Plenty of fruit trees on the estate, and we could sleep two pilots to a room, on great soft beds. A pretty chapel graced the inside of the villa. But there was fierce work to be done. The American Forces had just landed at Salerno beachhead, 15 miles south of Naples.   And they were in trouble as the German Artillery had secured positions in the walls of rock, 800 ft. above the invading Americans, and they were shooting down on our men on the sand beaches. We were not in their invasion plans as we were taking our mission assignments from the British, on the Adriatic side of Italy. The American 33rd and the 325th Fighter Groups were near Salerno for air support. But many infantry lives were being lost. So when Gen. Doolittle called, we geared up to help.

Our crew chiefs began fitting gasoline belly tanks under our P-40s and two 40 pounders on each wing tip. I was Duke's wing man for the first mission. He also was the navigating plane for 48 planes, climbing out over the high mountains of Italy. We were told to swing over the Salerno harbor, and go no lower than 1,500 feet above the American Naval Vessels and the merchant supply ships, who "would shoot at anything that came near them," as they might not correctly recognize us. We wanted to come barreling in over the white caps and surprise the enemy, but we were restricted by orders.

We had to approach the beachhead over crowded ships, anchored in the bay, then make our descent as sitting ducks, to the enemy gun positions in the mountainous wall above the beach head. I was with the first four P-40s to go in line abreast, firing our guns all the way. I ducked as red flaring golf balls went by my canopy. For every explosive shell, one was a tracking red flare shell, so the source of their shells could be seen We tried to silence them before they got us. But they never stopped firing. Then a sudden jerk and I knew I was hit. I looked down at the wing, saw an 88-mm shell hole in the leading front edge of the right wing. I pulled up abruptly, looked backwards for fire, or spraying gas, saw none, and headed to an emergency field to check out the damage. Hydraulic landing gear lines, control wires, and gas lines in the wing were not damaged. But it was a huge relief to find the small strip and have it checked out. The

88-mm hole was in a 12-inch square riveted section. That is probably why the shell did not explode. I took off over the mountains, singing Psalm 27 all the way home. Thank you, Lord, for saving my soul, in more ways than one.

The Germans must have been pulling out again, as we could see hundreds of trucks and equipment on certain roads, presenting targets for more strafing missions everyday. When I landed from a Sept 16th mission, I was just getting out of my plane, when I was greeted by Jack Benny and Winnie Shaw, who had just taxied up near me. They ate dinner at our officers' tent, had the usual spam, powdered potatoes and canned string beans, and then they put on a great Jack Benny type show for all the enlisted men and officers of the three squadrons that composed the 79th Fighter Group. Evenings around here were pleasant. Pomegranate and orange orchards provide pretty walking areas, even glimpses of the Mediterranean from the villa property.

Those strafing runs were causing some sleepless nights though. Got up in the dark for another mission to help extricate the American invasion troops near Salerno. Twenty-four of us flew to the target, dropped our belly tanks, and peeled off to drop our 40-lb. wing bombs on tanks and fire our guns at trucks. Schaffer was my wing man and followed me down. After the release, he called on the radio, saying Davis had been hit by the anti-aircraft fire and was already in his parachute. Duke Urich radioed me to find the nearest emergency field, to start a rescue search by our ground people, in order to find Davis. The next day, a Focke Wolfe 190 dove right in front of me at high speed. So fast we did not have time to drop our gasoline belly tanks and chase him, and I saw no bullet holes in my plane after landing.

Orders came through to pack up and move up closer to the bomb line, behind Montgomery's 8th Army. We would no longer be needed over there on the Salerno and Naples invasion front, and would resume support for the 8th Army, now moving toward Foggia by way of the eastern Adriatic shorelines. Two missions, and two nights later, we moved again to Picassa, 40 miles from Taranto. I rode in a Jeep, on dusty, truck crowded, roads. We passed hundreds of Italian soldiers walking the roads and railroad tracks, in their journey home to their families. Some had civilian clothes, others still Italian uniforms. As we got near our new airfield, a huge explosion went off behind us. A British truck had gone a little off the road and hit a land mine. Our new airfield, looked pretty good for a dirt runway. We were getting a bunch of new pilots to train. We had 11 and more were coming. The 8th Army had captured Foggia, so we expected to move there in a couple of days, when they had cleaned up an airport.

Foggia #6 was our new base. It was one of many airfields used by the Luftwaffe and Italians. Foggia was also a huge railroad center. We could see the heavy damage the big bombers did to all those railroad tracks that converge around the area. We waited for our support trucks to arrive with bombs, ammunition, etc., plus our meteorology crew, mechanics, and cooks with food. The next day, I was posted on the board to be the squadron flight leader.   Meaning that my job was to navigate our squadron's 12 planes to the target, and be the first to down into enemy fire.   In the past I was usually the wing man for our squadron commander, who always was our flight leader. I got them there okay, but my wing guns jammed on the way in, and as I pulled up I saw a large explosion behind me. Then a parachute, obviously, one of our guys (Capt. Neilsen) who was seen to land safely, but in enemy hands. My flight formed again, under heavy rounds of fire, and we headed for home.   Strafing missions continued each day as the Germans were bringing up all kinds of trucks and artillery to make a strong stand. We had lost several more pilots and planes that week, and we rapidly trained the new boys to take their place.

MONTGOMEREY'S DC-3 and STORM

Montgomery had made a friendly bet with Eisenhower, that Monty's 8th Army could take Sfax by April 15th. At that time, it was quite unlikely, so Ike's staff agreed to give Monty "anything you would like", if he could pull it off. When the 8th Army actually beat that date, Montgomery reminded Ike of the "bet". Soon he received from Ike an armed B-17 and an American crew. This enabled Monty to get to meetings from Cairo to Casablanca, during the North African campaign. But here in Italy, Maj. Benson flew into our field with a new DC-3 transport one day. He was a friend of Duke Urick, and asked Duke if he could park his DC-3 at our end of the field for Monty's use, replacing the B-17. It would be convenient to go to Algiers or Cairo, as needed. Major Benson was included in the gift package, so we saw a lot of him.. Usually, Monty drove in on his motor bike, from his headquarters, sometimes in short pants and sleeves, his trademark. Then he stood under the wing for shade, while the plane was getting readied. I never had the courage to talk to him, but we stood within 10 feet of him, sometimes shooting our cameras, as he smiled for us.

On Oct. 6th, we noticed that our target maps, put up daily, showed that German Forces were on an offensive and were pushing the 8th Army back. They called us for help, and between the RAF 239th Group and our 79th, there were 77 trucks and tanks destroyed. General Montgomery sent us a congratulatory message of thanks for our effort. We knew of at least two pilots who were lost, though, one being a famous RAF Squadron leader, Jackie Darwin. On Oct 8th, we moved to Foggia #3, a better air drome, 25 miles from the front lines. Yesterday, Rogers, Hundlley, Kirsch, Dzamba and Dietrich got orders to the States. They had flown on 80 missions. Mac and Adair were sent home due to Fatigue, and some high honors for their 75 missions. Only Hennin, Watkins, Lee, Owen and Morrison, Duke and I still remained of the original pilots. On Oct 10th, 12 of us were awakened at 5AM for a 6:00 o'clock take off. to Pescara, 150 miles behind the lines. We found the target, a long steam engine freight train. Our first run, dropping our 500 pounders, hit some tracks and freight cars, then we saw a huge explosion of steam, from the destroyed engine. Gen. Montgomery came by us that day, and was in a good mood, joking and talking before take off. The next day the 8th Army asked for some support in their attempt to capture a small town near the German and 8th Army front lines. The plan was for the British to shoot smoke shells into the targeted town, and we would drop our bombs in that zone. It worked well, and became the first of many future cooperative missions with the Infantry of the 8th Army.

Capt. Neilson, who had been missing in action 3 weeks ago, returned to our squadron today. He told us how he had bailed out behind the lines and was seen in an open field by some civilian Italians. He knew the German anti-aircraft batteries were firing at us in the air and had become so distracted that they missed locating him. The Italian farmers hid Neilson under bales of hay, fed him and clothed him. Then they told Neilson there was another "prisoner" they were caring for 100 yards away. Neilson met the British paratrooper, who had been captured by the Italians before their surrender. But now that he was free, with the Germans stationed all around, he had to lay low. The two escaped, by walking 25 miles at night time and got through the lines to Montgomery's troops, and then here.

Another request came from the 8th Army to destroy a German occupied town. Twelve of us effectively dive bombed the village, and we were promptly commended by the Canadian troops for the accuracy of the effort. When we landed from this mission, Gen. Doolittle had arrived and the whole squadron was preparing for an awards formation. Duke was awarded the Distinguished Flying Cross, Hennin and Watkins received Purple Hearts. Combat pilots receive an Air Medal for every 20 missions over enemy territory. At this point I had received three Air Medals, no big deal. You actually received one medal, the other two were signified as Oak Leaf clusters for attachment. The next day another pilot who had been shot down, arrived back. Lt. Davis, a neat guy. He told us that poor but friendly Italians had hid him in their home. One of the members had lived in New Jersey for 7 years. They nursed his broken collarbone and shoulder, and got him over the lines, so he could get back to North Africa for hospital care. Since we all thought he was a POW, his clothes, etc. had been passed around. So he had to retrieve all his stuff from his "friends". I got a surprise one day when my foot locker arrived on the mail truck, 10 months late.

Oct 25th. Today's mission was a large 60-plane dive-bombing run on a town up in the mountains. It was our squadron's turn to lead this one, and since I was scheduled to be the leader our squadron, that made me the navigating leader of all five squadrons. We took off four abreast, rendezvoused with the other squadrons, and headed towards the target, 80 miles away. I took the proper magnetic compass course, compensating for the high winds, and magnetic deviation. When we arrived at the target, I knew we had the right place, but a radio voice clearly said " Do not bomb Spinette any more, bomb Bojano". While we were coming closer to our peel off, the message was repeated. Then I heard one of our pilots yell, "Blow it out your barracks bag." I gave the Tally ho order and led them all into steep dive bombing runs on the assigned target, and we formed up and went home. At our mission briefing we all concluded the Germans had a high frequency radio from one of our captured planes, and had tried to spoof us. More missions on bridges were required. Bridges were tough to hit directly, but enough bombs got the job done.

On Oct 29th, I was on the morning mission up the Adriatic Coast looking for Enemy ships in small harbors. We returned for a needed rest, but about 3:30, I was asked to fill in for someone on the 4PM mission. I hesitated as there were gathering storm clouds, but the meteorology people said we would be OK. Going out, the cloud formations looked ominous. After our dive bomb run, all 16 of us, 12 P-40s and 4 RAF Spitfires, hugged close to the water, as we headed home. But black, fierce clouds, and lightning up to 20, 000 feet, hemmed us in. We normally didn't talk on the radio, but this demanded a conference. Should we look for a beach, a flat farmland, and belly land to at least save our lives? We did not have enough gas to go around or above this thunderhead. We talked on our radios, about the options. Then it dawned on me. I radioed our pilots, " you guys stay right here in a circle, while I take a 270-degree heading inland" . I signaled my wing man to follow me to something I had seen on the morning mission. Time was running out, because the storm was getting closer. Within 5 minutes there it was, a freshly plowed landing strip, which was still being prepared. We buzzed it at 15 feet to check the condition. We also saw one tent, and a plow. I radioed our squadron to head 270 degrees, and join us, as I was going in, wheels down, to make sure we all could get in safely. Sure enough, I made it okay on the unfinished dirt strip, and 12 P-40s and 4 RAF Spitfires followed me in to land safely. We taxied over to the tent. The tractor man, fortunately, was not a German, but a British workman, preparing an emergency field for a future day for RAF and American boys. Lt. Jaslow, happily climbed all over me with joy to be on the ground, as it was his last mission in combat. Within 5 minutes, the storm broke on the little tent with lightning, wind and hail. Thank you , Lord. What a day.   The workman got a call through to his headquarters, and the message was that trucks would be arriving about 11 PM. We got back to our beds about 3:00AM and returned two days later to fly our planes off the dried-out strip.

Nov. 1st. More close support targets, except this time we carried two 250-lb. bombs, and would do both dive bombing and escorting some A-20 bombers. We got to the target and watched the bombers gather closely in a tight formation as they readied to release their load of bombs in a tight pattern. The 88-mm shells were exploding all around, with big black puffs of smoke and steel. Then one A-20 got a direct hit and exploded. All we saw that remained, was two parachutes, and a second A-20 in a steep spin to the earth, out of control. The rest of the bombers dropped their bombs and droned on home. Boy, was I glad I was in a fighter and not bombers. Thank you, Lord. The next day a squadron of P-38s landed at our field and stayed overnight. The P-38 was a long range two- engine fighter (one pilot), and they were scheduled to rendezvous with a huge B-24 and B-17 Bomb Group, and escort them for a bombing run on a Messerschmitt factory in Austria. We learned they were attacked by 125 German fighters and ferocious 88-mm anti – aircraft fire, coming and going. One B-17 arrived back at our field with holes all over it and with a gunner who had been dead for two hours. The Luftwaffe did not have that number of aircraft to use in our theatre as they were just starting to save them for their homeland defense and the Russian advance. To assist the 8th Army get a stronghold in their push up the Eastern coast of Italy, three of our squadrons took off early. Our planes were all over the sky, when I suddenly saw a string of Focke Wolfs in a dive-bombing run on British trucks and tanks below. They were heading for home, as George Lee and his wing man dove down on the deck to try and catch up to them. I followed at higher altitude, but we soon lost them in the haze. All we got was friendly ack-ack, which was not so friendly.

80TH MISSION - CAPRI AND HOME

The next day, I got word from our Flight Surgeon and our top command that I had flown 80 combat missions over enemy territory, would be released from combat flying for return to the U.S.   Orders would come in a week.   Also, our squadron received a congratulatory message from the 8th Army headquarters saying, " the accurate and concentrated bombing in the past few days has helped considerably to accomplish our objectives" . Of even more importance, our 79th Fighter Group, received a Presidential Citation from Washington, D. C., signed by four- star General Arnold, Chief of all military operations. This would permit us to wear proudly on our dress uniform, the beautiful gold bordered blue pin.

Before leaving for home, Hennin, Jaslow and I drove a Jeep over the mountains to the Isle of Capri. The Air Force had made most of the hotels available to pilots for rest periods. The first night we stayed in Amalfi, where Longfellow wrote " Amalfi by the Sea". Nice to have a good meal while violins were playing. The next day we took a small boat to Capri, and enjoyed the Italian atmosphere. You would not think there had been a war all around them a month ago, as Salerno and Naples were only 8 miles away. We took the cable car back down to Grand Marina Harbor. When the tide was at low point, we took a small tour boat to the cave entrance of the Blue Grotto. The boat barely squeezed in that hole in the cliff, but inside it was a beautiful sight of blue water, caused by the sunlight shining on the 40 foot deep blue rocks. We spent a day at the foot of Mt Vesuvius, in order to see the famous ruins of Pompeii. The guide told us Pompeii had been covered with over 15 feet of pumice and ash from the epical eruption in the year 7 AD.   Gases killed everyone, but the beautiful marble buildings and streets were all in tact when the ash was removed. The Roman columns all over, however, were explicitly decorated with male genitals. Was this a clue to its destruction? Another Sodom and Gomorrah?

When we got back, our orders had come through, simply stating to return to the U.S. at the best possible way. So the next DC 3 that was heading to Africa, took Watkin, Hennin, Jaslow and I, stopping at Naples, Tunis, Algiers, Constantine and finally Casablanca on the Moroccan coast. There we could sign in for the next troop ship to leave for the states. But since there was no ship leaving this week, we decided to go over to Marrakech, about 150 miles east in Morocco, at the foothills of the high ranged Atlas mountains. We thought it would be nice to see a little of England, and go home from there. The plan was to stay out near the airport, get up at 3 AM each morning, so we could talk to Pilots leaving for England with their Bomber crews. We begged a ride with them, but to no avail. Each crew was loaded to the hilt with supplies and men, just coming over from the states, and heading for combat duty in England for the invasion of Europe. They would leave here, fly up the coast toward Gibraltar, then turn out to sea 100 miles, out of range of German fighter planes operating from Spain and France, until they came to England. So we gave up after 3 morning tries, and returned to Casablanca to board the converted troop ship, the Empress of Japan, which was headed for Newport News, Va. We thought it pure luxury, five beds in a stateroom for Officers. The first thing we did was soak in a completely filled hot bath tub. The ocean trip was pretty calm, but Lt Jaslow seemed to be sickish the whole voyage, getting weaker every day.

Arriving at Newport News Va., we docked and checked in at the military base, Jaslow looked so bad I said "You need a hospital, not a train home to N.Y." to which he agreed. Except the hospital was almost 2 miles on the other side of the base. We both spotted a shiny new open Jeep, right in front of us. We decided to " borrow" it, after all, we had owned the skies, why shouldn't we use this for a little while? Since the key wasn't there, we got out a large 50 cent piece, which Jaslow placed under the dash and held it over the key terminals while I started it and headed for the hospital. Parked it right at the front door. The verdict of the doctors was that he had jaundice and would not leave for home till 5 weeks. But before saying good bye he handed me about 10 rolls of Kodak film and asked me to get them developed and mail them to his home. I assured him I would, but I would like to make a copy of them for myself. What a break. Those pictures of our year together comprised my treasured album. When I walked out of that hospital I never touched that Jeep again as the better part of valor, and was content to walk back, and let an MP return that Jeep to the angry unknown Colonel.

A telegram was sent informing my parents that I was taking a train to Penn Station in NYC. Telephones were too expensive then. But when I got there, no one had arrived. I walked up and down the station for over an hour, then decided to hop a bus for home. But nobody there, either. I raided my mom's refrigerator till they burst in the house with all kinds of stories to tell about the delays of the evening. Great to be with my folks. It was 2 1/2 years since I had been home. After a few days home I had to leave for the Ritz Carlton hotel in Atlantic City, for Pilots rest and restoration. I enjoyed the luxury there , but I also liked the fellowship of the Baptist Church, to which several of us went. After a leave home, I was ordered to a P-47 fighter base at Providence R.I. We would learn to fly the Thunderbolt soon. But before I did that, I was offered an opportunity to go off to the Air Corps Intelligence School in Harrisburg. I bought a used 1937 Plymouth, my first car, and spent 2 months at what had been an elite boy's school, on the Allegheny river.   After graduation, my classmates were sent to mission intelligence assignments, but they did not know where to send me. The problem was I was the first pilot to have graduated from this school. They finally ruled that a pilot could not be an intelligence officer in any squadron. That was a ground officer's job and flying pay would not be allowed. I had the choice to give up flying, or go back to Providence to fly, and train others in combat tactics. I chose the latter.

So it was back to Providence RI and check out in the P-47. It took a few more days to study the cockpit and take off and landing procedures. However, the office reminded me that there were only 2 days left to the end of the month in which I had to get 12 hours flying time on my record. To get flying pay, it is necessary that a pilot can not let 3 months go by, without flying. Since a minimum of 4 hours airtime per month was required, I had to get in the air for 12 hours, in the next 2 days. Because the

P-47 used 1 gal of 100 octane gas per minute, they gave me a Piper Cub. That was awful. About all I could do to fill in the time was to shoot landings. The Piper Cub was not designed for severe dives, climbs, loops or Immelmans, so all I could do was straight and level, bouncing in the sky over warm thermals. I got the 12 boring hours, and the flying pay. That completed, I made my first solo in a P-47. This is a great plane to fly. I wished that we had them available for us in our combat theatre of war. Maybe my boys out there will be getting them soon.

Back in the states with a P-47.

MEETING GWEN

I was home for a weekend, and went with my parents to the Yonkers Chapel annual spring Bible conference. The elders asked me to give a report of my recent overseas experiences, so I told them how the Lord was my strength and my salvation, and what an impact Psalms 27 was on a daily basis. I also mentioned one incident when Duke and I had had a very close shave. He said to me ."Lydia, you must-----"and before I could finish the sentence, the audience burst out in a roar of laughter. All the old timers there knew of Lydia Pinkham's pills, a medicine for women. At refreshment time, a lovely 21-year-old came by with a tray of sandwiches. She was so attractive and nice I wanted to talk with her some more. Later, I was invited to stay with some young people for games, and there she was: Gwen Dunkerton. I had known her family over the years, and she knew of me through my cousin, Lois, while they were both at Bob Jones College. Before we left that night, I had a date to meet her in NYC the next weekend. And then more dates and even a dinner at her mom's house. One Sunday afternoon, at about 2 PM, I called Gwen to tell her to get out-side the house at about 3 PM. I would be buzzing their house in a P-47 Thunderbolt, one of the best fighters of the war. The whole family was there, Tom, June, Joanne, her mom and dad. I made a few low passes and headed back before running out of gas. After a lot of letters, telephoning, dinner dating, going to some Jack Wyrtzen gospel rallies, meeting our families, etc., I got up the courage to ask her to marry me. And soon after she accepted, I got orders to be transferred to a fighter training base at Dover Delaware. That was at the end of the world to me then. Since I had a four-day delay en route to see Gwen and visit home, I put on my full dress uniform and headed to the 2nd Air Force Headquarters in Brooklyn. It was from here that ultimately, my orders had come. I started with a Lieutenant, told him my dilemma, to which he referred me to a Captain, who then referred me to a Major, who finally let me talk to the top Colonel.   I told him of our engagement, and asked if there was someplace else in his command that I could be used, besides the one at Dover, Delaware. I think he must have been a Christian, he was so helpful. He said, "Go home for a day or so, and wait for an answer by telegram". Sure enough, the telegram came stating I should report to Riverhead L.I. at Suffolk Army Air Base, a P-47 for training combat tactics. Another gracious answer to prayer. Thank you Lord. What a difference that would make. We were able to meet in N.Y. for special events, and her mom even gave me the maid's bedroom in their big home at Rumsey Rd. some weekends. Her mom even let Gwen spend a weekend with me out at Riverhead. I found a nice hotel for her, and we had a fantastic time at the Southampton beaches, and dining places. The trusty old Plymouth made other weekends and holidays possible together for all kinds of picnics and family outings.

1944

WEDDING & GWEN AT LONG ISLAND AIR BASE- WAR ENDS

Oct 14, 1944 was our wedding at Bethany Chapel in Yonkers. Uncle Richard Hill married us, as he was Gwen's choice and Uncle John also officiated, as he was my favorite. Both were Brethren preachers. After an exciting reception for a crowd of family and friends at Gwen's home, we drove to the Hotel Thayer at West Point. We attended the Sunday Cadet Chapel's service and headed to Split Rock Lodge, on limited gas rations. After our honeymoon we returned to our nice new apartment. But when I returned to the squadron, I found all pilots on an alert. They were flying P-47 patrol missions from Montauk Point almost to Coney Island, searching for possible buzz bombs that might be launched from a German Submarine. So I was out several times a day, and into the night, cruising at 7,000 feet, and searching for buzz bombs that would be cruising at 4,000 ft. altitude and 350 mph. U.S intelligence had word of a surprise attack on NYC. After two weeks it was called off and we resumed our normal training of the experienced pilots who had come to us for combat tactics, high-altitude flying, night flying, ground gunnery, and air gunnery. The pilots we trained flew their own planes while they followed leaders like me, in the sky. Sometimes we ended up near NYC, so I took them over all the sights up the Hudson river. When President Roosevelt died suddenly, the burial service was to be at Hyde Park, NY., his home. Our outfit was asked to do an appropriate fly over for the occasion. I flew on several formations over Manhattan, for special Victory Bond drives that sometimes were difficult, due to haze over NY, and the 1,500 foot requirement, over the tall buildings.

One day, the Operations Officer asked me if I would spend a week with a photography outfit. They had a B-25 bomber in which cameras were placed in the side of the fuselage, in order to get movies of sharp turns, dives and other photographical maneuvers, to be used in a training film of some sort. All went well until the day of a big thunderhead storm. It was far west of us while they were shooting our maneuvers, but soon I realized the B-25 camera plane was heading right into the black stuff. I knew it would be turbulent and dangerous. I noted that my partner P-47 plane, Don Tunis, who was following the camera plane, continued as though he didn't see the danger. So I radioed Tunis and said "I am getting out of here," then pulled away and landed. Later, Tunis related that soon after I called he was violently flipped over on his back, and then put into a dive, out of control, and barely pulled out, down at 2,000 feet. No fun inside those thunderheads.

It was nice to have visitors drop in out there. Mac Peebles, now a Captain in the infantry, came by, on his way to Tennessee. And shortly after that Crow and Jeanette Stahl called us to invite us to dinner at the famous Hotel Astor on Broadway. They wanted to celebrate the war ending, and the survival of both of us in it. Crow had been a B-26 Bomber pilot. Also Dad, and my brother Manny, came out for a visit, and it was possible to give each of them rides in the back seat of a BT-13 military plane. I gave my brother some acrobatics, even a slow roll, but didn't want to torture my dad that way. Gwen and I lived like Kings in a cottage in Hampton Bays, one block from the ocean, and didn't realize how well off we were. And like most men in uniform then, we just couldn't wait to be a civilian again. So when both Germany and Japan surrendered, we were given an opportunity to be discharged, based on a point system. Pilots were then "a dime a dozen" and the highest paid, so were on the early lists. However, before we left the base, we all took special exams and filled out required forms for the FAA to give us a Civilian license for flying airplanes up to the engine rating of our P-47's. It included an instrument rating, as well. Thanks, but no thanks for any more flying in planes like that Piper Cub, so I had no further inclination to fly. And beside, it was so thrilling to have flown all around NYC for more than a year. Freely buzzing my sisters' houses in a Thunderbolt, a feat that I would never be allowed to do again, even if I had to pay for it.

FAMILY HERITAGE

When in civilian clothes, there was a feeling of freedom. A good sense that another life, absent of war, was ahead of us, and we prayed that the Lord would lead us. Ephesians 2:10. Gwen and I wanted to be near our parents. We were ready to settle down and raise our own family, modeled after our godly parents. It has been said that we are who we are because of our heredity, which is 50% and our environment, which is the other 50%. I am thankful for both our parents' heredity, and the quality of life they lived. They demonstrated Christian values to us and gently encouraged us to follow the same pathways. At dinner and lunch meals, we were expected to be at the table on time. In the 1920's, the Bible was read by Dad, and then we all got on our knees, at our chairs, to pray. Then we could leave the table, but not before. At church time, all the children went with our parents. We were blessed with that kind of heritage.

My mother, Alice Elizabeth Pinkham, was born in Manchester, England, in 1884. When she was 8 years old, she was on a ship leaving for New York. The Ship's manifest and immigration records at Ellis Island confirm the sailing of the Hill family in 1892. Missing on those records was two of her brothers and her father Richard Hill. He was an evangelist who preached in Brethren churches all over England and Scotland, and occasionally in the U.S. He died shortly after the family moved to Jersey City, a country town across from NYC. So the widowed Catherine was alone with her 4 boys and 4 girls in a strange country. In her later years, she resided in our home for about 15 years. The story that circulated was that Grandma was a trainee "Lady in Waiting" for Queen Victoria. I always thought Mom treated her like royalty. Later it was in Ridgefield Park that Alice, Eleanor and Mary moved next to each other. Phoebe, a nurse missionary to Santa Domingo had married Asa Moore, who was a fine Baptist minister for many years. I learned some life-changing values in their home, spending holidays with Cousins Eunice and Lois, with whom I remained close friends.

In 1905 mom and dad bought their new four bedroom home (1 bath) next door to Mom's sisters. It cost only $4,000 and dad got a $4,000 mortgage to pay for it. Next door were the Swans, and immediately behind us were the Gilchers. In 1905 those houses were heated by hot air from a coal furnace, the lighting was gas piping to each room with a mantle, a wick like glowing device in the ceiling or wall lamps. The early ovens and stoves were gas. In the basement, the washing machine was just two big sink tubs, with hand cranked rollers that squeezed enough water from the clothes to be able to hang them. Mom used a long pulley line to the post on the property edge. Later, about 1927 or 1928, electricity became available, so my dad wired the whole house, by putting the wires through the gas pipes. And that amazed us. Mother was a great cook. She believed in cooking all the wholesome things, like turnips, spinach, and brown rice pudding. On Sundays, she always cooked a roast beef that she bought from Mrs. Kern's grocery, out of her meager weekly food allowance. She bought only whole wheat products from the Dugan man, who came to the back door. Occasionally, the fish man would come down our street blowing his long tin horn until mom came out and yelled " Mackerel" to him. It was a big thing to run out and watch him make the fish scales fly as he attacked it with a vicious long knife. I remember the day, during the Depression, a man knocked on our back door, and asked for something to eat. Mom told him to wait while she made him a whole wheat sandwich, which he ate while sitting on the porch stairs. We had some wonderful meals on the big holidays. A big turkey, creamed onions, mashed potatoes and at Christmas, homemade plum pudding with the rich butter sauce Mom called hard sauce. Always a feast.

My mom had four brothers, all Preachers. The oldest was James George Hill, a Baptist minister in Cazenovia, N.Y. The other three were Ministers in Brethren circles. Uncle John, whose wife was Aunt Lizzy, was an invalid, lived in Hasbrouck Heights. He was a most sought after Conference speaker and a super Bible teacher, in the north Jersey area. He would take Brethren church groups to the Planetarium in NYC and explain the wonders of the Heavens and relate it to the scriptures. He also preached regularly to Wall Street business men on their lunch hour, in the Historical Cathedral downtown on Broadway and Wall Street. Uncle Richard did the same from Sea Cliff, L. I. Where he ministered and organized bible conferences. He was the most sought after marrying minister among the NY- NJ Brethren. In fact, he officiated at our own wedding. Lots of brides in the N Y area wanted Uncle Richard to marry them. He had been a missionary to Persia a short time but had to leave the area. In the twenties, he started a missionary training center in Brooklyn that served as orientation for most of the Brethren missionaries in the twenties. Because all missionaries left on ships from the N Y harbor those days, the school was a greatly used place. Rowland, the youngest of the Hills, served in India as a missionary for 30 years. So we only saw him (and Aunt Diana) every 5 years, on furloughs. He lived to 102, and like John and Richard, was self taught in the Scriptures, but highly educated. He wrote a lot of Christian literature, pamphlets, tracts, etc. for the Bangalore, India people. He was a story teller, and I'll never forget his account of the 8 foot Boa that came up out of a toilet hole in the floor of the bathroom, while Aunt Diana was naked in the shower. He used a forked shaped stick to hold the snakes head while Aunt Di ran out of the room.

My dad was Harold Walter Pinkham, and came from a large family too, except that none of his siblings ever came to America. His father, Richard, came often to N Y, on his own ship, as Captain. That was before 1905, and he actually died in NYC, where he is buried at the Woodlawn Cemetery, in the Bronx. On one of those trips to N Y, he dropped off two sons, my dad and his brother, Norman. This was not unusual to my dad, as before he was17, he had sailed around the world twice with his father. He was tutored by a ship mate or his father. One of the souvenirs he kept was a 3-foot long photograph of Hong Kong harbor, which never ceased to intrigue us. It hung on our dining room wall, and I recall many moments examining the three seams of the pictures that composed the Hong Kong of 1900. Both Norman and dad accepted Christ at a brethren Chapel in the upper Bronx, where they started very fruitful Christian lives. Norman died a couple years later, at 24, and left a very lovely fiancé', named Gertrude, who continued close to our family, over the years. We all called her "Aunt Gertrude".

Dad did not get back to England for 30 years, until 1933 when he lost his job during the massive Depression. I was 15 then and remember seeing him go off on the German ship *Bremen*, and returning on the *Europa*, out of New York harbor. It was a whole month's reunion with his family, and an emotional time, as there was just one sister in the family of seven boys. Dad loved Elsie and corresponded with her. The other brothers joined up from different parts of the British Empire. Carl came from Italy, where he was a missionary in Foggia. Edwin came from Australia. Cyprian lived in England then, and was involved in an anti-war movement. And I think Herbert, a ship captain, was there, too. The spark plug of them all was Courtney, an architect whose lovely home was in Bristol.

On his trip, Dad spent a lot of time with Elsie, who ran a boarding school for girls in Swansea, near Mumbles on the Wales southern coast. She was fine looking lady who never married. My brother Mansfield, visited Mumbles area in the late 1980s and happened to run into a lady who grew up at Aunt Elsie's school. She wrote Manny about some lovely memories of her training, and sent some fine pictures of those days, which she wanted us to have. I can understand now, why Dad was so fond of Aunt Elsie. Courtney was the youngest brother, who gradually became the only surviving member of the whole family. He wrote long letters to Dad, and even wrote occasionally to me, while I was in college. During the war, when he found out that I was flying in support of the British 8th Army, he sent me a very detailed ink drawing, with beautiful designs. He titled it with large colored letters, JACK THE GIANT KILLER. At the time, of course, I thought that quite ridiculous. Happily, Gwen and I met his widow, Aunt Esme, while visiting Bristol. Dad's job in New York City required a commute every day on the steam engine train that ran through the dark smoky tunnel to Weehawken. Then a 20-minute ferry on the Hudson River to about where the Twin Towers stood. Sometimes on a Saturday, he took me to his office building. I spent the whole morning riding the elevators, or walking down to the Battery to watch the fish, in what was the largest ocean fish tank in the U.S., the Aquarium. Sometimes, he would drive us out to a Patterson park, where he could watch a cricket game, or even a little rugby, something to remind him of his English home.

Our Ridgefield Park home was on Lincoln Avenue, across the street from Lincoln grammar school. Behind the school was our constant play ground, where we could always get into a game. Depending on the season, we played either marbles, baseball, basketball, or touch football. We learned how to play from each other. No coach, so when we chose up sides to pick a team, one had to suffer the humiliation of being last pick. And you learned to improve that way, as some day you might be selected as the pitcher, or some other key spot on your friend's team. Those days, grades and reading were set aside. It was always more fun to be behind the school, competing with any of the kids that would show up. Most of the houses nearby had front porches, and ours had a nice two-seater swing, hung by two chains. On the railing grew a massive honeysuckle, where my mother put the canary cage and occasionally, the parrot cage. From the swing we watched the Model T Ford's go by.

Sometimes it was Mrs. Doze's drunken son, looking like Abe Lincoln, wobbling. home from the town "speakeasy". Once he met a friend in the same condition, across the street. As they passed each other, the friend suddenly hauled off and knocked out John Doze cold and flat on the sidewalk. His mother rushed out, got John up on his feet, and was so embarrassed she never spoke a word of the incident again. Also, the bravo kids, precariously climbed on the brick wall ledges of the school building. And occasionally, some guy got up on the long slate roof, walked up and down it, until our nosy neighbor, Mrs. Doze, called the police. The candy store on the corner, featured penny candies in a period when a cent would buy a bubble gum, or a chocolate cream, and five cents would buy a Pepsi. I never got an allowance and neither did my older brother, Mansfield.

Mansfield earned his spending money delivering newspapers in the early morning hours before dawn. And sometimes he got me up to help him fold them just so, in order to pitch them to the porches from his moving bicycle. Occasionally, he came home with a little red wagon with a missing wheel, or a discarded bicycle from someone's trash pile. But always on Sunday mornings he brought home the colored funny sheets. They were special, because my parents never bought a paper on the Lord's day. There weren't enough bedrooms for all of us, so Manny and I had to sleep in the same bed. He had the notion those days that a window had to be wide open. It was great for fresh air, but it seemed we all got sore throats. Quinsy sore throats, the kind that's like strep is today . We had only aspirin to treat it, as penicillin was not known.

Dad gave us jobs to do, while he was at work in New York City, like shoveling coal into the basement furnace. Then we had to shake the ash from the grates, and put the ash cans out for pick up. We helped to paint the house occasionally, but cutting the grass was a regular chore. I enjoyed keeping goldfish, but the only place in the back yard for the fish, was in an old cast iron porcelain bathtub, dug under the apple tree. The fish couldn't tolerate those wormy red apples when they contaminated the water, and died of asphyxiation.

Both of my sisters, Alicia and Eunice, have lived to 96. Lee, the oldest, was a graduate of a cooking school, and also a very fine pianist. She was a great accompanist and a good church hymn player. She was a great encourager, too, and I always enjoyed being with her, and now miss her. Eunice, who now in 2007 is 96, still dresses up to look like a million dollars. She has all the charm, and at one time became the President of the Red Cross for New Jersey. I still recall her single days when she worked in New York for Collins and Aikman. I guess the lingering memory is there, for when Aunt Annie moved out, Eunice got the front bedroom because she was a good breadwinner. When she married her Tom (Munro), they bought a lovely summer home at Lake Mohawk, N.J. and entertained friends and family very often. I learned to sail there, on their Snipe, the same sailboat, their son Bruce sailed when he won the National Presidential race on the Potomac. At 96, Eunice is still driving, still flying and still proud of both her sons' family accomplishments. Eunice is our model for longevity.

Gertrude, the youngest, was born on my birthday, nine years later. Gertrude and I have been close to each other, and I have enjoyed her as a good friend, and she is nice to be with. During the Depression of the 1930s, when my dad lost his job and was out of work for a year in 1933, Gert and I went to Camp of the Woods with Mom. The same year my Aunt Annie moved into our house to help pay for the expenses. When Mom was widowed by Dad's home going, Gert was still at home and was a special help to her. She was a successful business woman, an executive assistant in a large Jersey bank. She showed great endurance, when she lost not only her husband, but also her son Richard, who was only 39. He had trusted Christ in his last year before he went to be with the Lord, and for Gertrude that was comforting to know. She is now a star retirement community leader.

Very seldom did all the Hill families get together. But the big one was in our back yard, for Grandma Hill's 90th birthday. Maybe 40 relatives attended. We still have some great movies of that time. The Swans next door, were the godly parents of Colvin, Catherine, and Eleanor. As I grew up, Colvin and Catherine had already moved out of their house next door, leaving only Eleanor, who got into some pretty bad company, and literally broke her parents' hearts.

Uncle George remarried a lovely English nurse, Phoebe. And the Gilcher sisters, continued at home while commuting to their secretarial jobs in New York. They married the finest of Christian men. We thought it was great when those cousins bought brand new black Model T 's or Chevy's with a rumble seat. Richard also worked in New York, but his heart was always in the garden, and he believed in growing things organically. Phoebe practiced hard on their piano and played some difficult music, like "March Militare",which we could hear from our house. Often, Annette and I played Parcheesi, or Flinch, and she was always so patient with me. Though I was four years her junior, she treated me as an equal, and I have been always grateful to her for that .She still is a fine person, that everyone loves. I admired her for her faithfulness to the Lord, and for her pretty brimmed hats she wore the days we walked to the train station. I still remember their kitchen table, full of figs, dates, fruit, and the wishbone from Sunday's turkey dinner. I'm afraid I bothered the daylights out of them when I was a kid, but they were always kind, even when I wistfully hung around to get a ride in the rumble seat of their new Model T.

As a teenager, I enjoyed working outside in the empty lot next door. It was full of weeds and was so shabby that anything I did to it made it look good. I took pride in that place when the weeds were cut to look like grass. But the most fun was the goldfish tank, really an old white porcelain bathtub. It was eventually moved to the base of the wormy apple tree, where the fish kept dying from the leaves and apples. When I was 15, I had my tonsils out. The anesthetist gave the ether on my mother's bed, and the doctor gave me the tonsils in a pickle jar. It was the same bed on which my mom had given birth to me. It was a surprise when Dad brought home a second hand trumpet, a beautiful Conn 22 B. I soon found a teacher, the father of a school mate, who had been a trumpet player under John Phillip Sousa. In high school, I was allowed in the band, and later in the orchestra. This gave me a chance to apply to Pop Tibbitts, for a full summer's job at Camp of the Woods, one of the biggest events in my life.

At 13, I knew there was a decision that had to be made as to which way to go in life. Was I going to follow the path of my school friends, without the Lord, and go in my own way? Or, would I yield to the Lord, let Him direct my paths, and then live for Him. I was baptized soon after I made the right decision for Christ, and started to grow in a greater knowledge of my Lord and Savior. At 16, I was teaching a Sunday School class. And when Bill Stephens, a dapper fellow and our Sunday School Superintendent, was moved away, that left me to take on his job. (He married the daughter of the Bethlehem Steel President and then he later became the President) We had only 35 or 40 kids from all over the town, but Annette was a faithful teacher, and Phoebe was always there to play the piano for the hymns. Things went smoothly until the father of one of the poorer kids died. The mother called me to handle the funeral as they had no other church contacts, other than our little Sunday School. My dad wisely encouraged me to go through with it, though I did enlist the help of Bob Fenty, a 30- year old male Sunday School teacher. I remember almost fainting in the kitchen of the deceased, since I was the first visit there to make arrangements!! Those last two high school years, I worked summers at Camp of the Woods.

At age 17 in 1935, I enrolled at New York University, and commuted from home by train, ferry and subway. I even got a job, from 2 to 5in the afternoons. Forty cents an hour helped pay for the text books, and all the egg creams I could drink at the subway cigar counters. By subway and busses, I delivered upholstery goods to New York hotels I had to measure the piece to cut it, then take it to the hotel housekeeper. Dad let me stop working, summers, so I could go to Camp of the Woods. Those four years of summer months, at Camp of the Woods, were like a college campus to me, though it meant living in tents. The great guys and gals there became my lifelong friends: Ted Carlson, a Yale student then, later my best man; Crow Stahl, then an Eastman School of Music trumpet major, later became an Air Corp B-26 Bomber Pilot; also Ed Taylor, in hospital management; George Sadler, popular as a Baptist pastor; and John Maltese, a great concert violinist with whom we have kept close.

After two years of college I switched to the night program. Union Carbide hired me as office boy at $65 per month and paid half of my tuition. I served the secretaries of the President and three Vice Presidents for those final three years and summers. I worked from 9 to 5 at 30 E. 42nd St., then took a subway ride to night classes from 6 to 10 PM. A raise to $75 a month came when I got my sheepskin in 1940. Fortunately, my dad worked three blocks away, as manager of a 20-story building at 8 West 40th St., so we frequently had lunches together. He was my best pal. My friends then were mostly guys and gals who had been to Camp of the Woods, because there were 13 from our town that had been counselors. This group started a Saturday night get together at our Chapel, and soon 50to 60 young people were enjoying great fellowship in music, singing hymns, choruses and Bible teaching for youth. Usually, on Friday nights, I took a subway from my office, to Yonkers, where I played my trumpet for the street meetings at Getty Square. I met some nice girls there too.

ON THE MOVE WITH EVEREADY

When the end of the war came in 1945, a return to my career in the sales department of Eveready batteries and Prestone, brought me right back to30 E. 42nd St., to the offices of Mr. Berdan and Mr.Kleinsmith, vice presidents. They wanted me to train in the sales division, but said they didn't know how much to pay me, since the post-war salary levels had not yet been established. I told them I had confidence they would be fair, and was ready to work, whatever the salary was. It was fair, and about what I had received as a pilot. We had to move in with Gwen's parents because of the scarcity of suitable rentals. 139 Rumsey Road, Yonkers was luxury, especially her mom's cooking and her dad's supply of good Iowa corn-fed beef. We enjoyed Bethany Chapel and the spiritual life at such a progressive Brethren Church. Mature elders were there, like Don Parker, Mr. Munroe, Claude Speicher, and Gwen's dad, Orrin Dunkerton. Don Parker was a mentor to me over the years, but more than that he was a great encourager, a builder of young people. He was so well thought of at Union Carbide that they employed many young men and women whom he recommended. At a national sales meeting a few years later, I heard Jack Spangler, the President, tell a few of us at a reception, that Don Parker was a man "who practiced what he preached," and that Spangler had just completed his new will in which he requested that Don Parker officiate at his funeral service. What a fine testimony.

Traveling by Pullman car to upstate New York, leaving Sunday evening and returning Friday, soon became a chore. Post-war airplane service was uncommon and even new autos were scarce. But management was watching the results in the Watertown, N.Y. area, where they had sent me, because it was a sales test of revolutionary synthetic motor oil, which Union Carbide had been selling in Alaska during the war. Advertising sold a lot of our new Prestone Oil then, but the detergent part of the oil created so many clogged engines, that the test was ended. But there was plenty of work selling Eveready products in Connecticut and New York to keep me nearer home. My dear dad, at age 62, suddenly went to be with the Lord. At my young age of 28, I thought he had lived a long full life, which he had. For years he was the song leader at major conferences of the Brethren; he also sang solos, was a faithful elder and leader, plus an active Gideon. A prince had fallen. My best friend and my dad. This left my mom and Gertrude, who was still single, at our home. Gertrude was good to Mom those days, and Manny, Eunice, and Lee stayed close to her, and provided for her in so many ways.

Eveready moved us to Rochester in 1946, shortly after Dad died. We bought our first home. Our affordable house was on a dirt street and cost only $10,000, $500 down for veterans. We fellowshipped at Congress Ave. Chapel, and aimed to serve the Lord there. Godly families were there, such as the Larters and the Westfields. And it was the home of two popular Brethren preachers, John Bramhall and Harold Harper. We enjoyed entertaining preachers on Sundays, and we even housed several returning missionaries, for a week or so at a time. They included men like George Rainey, and Mr.Tharpe, both of whom had been forced to escape from China when the communists took over after the war. Mr.Tharpe still had a big chunk of Chinese gorgonzola cheese, which was so smelly he it kept out on the window ledge on cold winter days. Barbara was born here and was our one and only, enabling Gwen to travel with me to places for Eveready all over the Finger Lakes region. Also it was very special when my mom came to visit for a week when Barbara was two and really cute.

But a bigger territory opened up in Buffalo, so we found a nice home about 2 miles from Kensington Chapel, the Brethren church of our choice. We knew some of the Christians there so we soon became involved . Patsy was born while there, as was John. We started a craft club, on Friday nights, with 60 kids making everything from ceramics to kites. Later as Sunday School Superintendent, we ran a contest in which fertilized eggs were awarded, and timed so they would hatch at Easter. The eggs were kept warm by a GE light bulb in our basement. Some never hatched so we bought enough one-day old chicks to satisfy the kids. Strangely, the mothers were not too happy with the new prize in their house. Another time Les Paulsen got a helium tank from his company (Linde) so we could blow up balloons , attach a gospel tract , and let each kid send off a balloon. We actually got a few responses.   The three Brethren churches in Buffalo had great conferences.   The downtown Assembly was known as the missionary-sending church. At least seven families went from there, some of whom were the Logans, the Dibbles, the Hortons, and the Brooks, all of whom had children who also continued as second- generation missionaries.

We spent our summer vacations at Camp of the Woods, usually going the same two weeks with Ethel and Vernon Larter, plus Bob and Cathy Harris. Bob and I made a neat aluminum sailboat, just for use there. Bob was a fine person, a good friend, who loved the Lord and the outdoors and could fix anything that needed fixing. We hunted deer every fall with shot guns, and even with bow and arrows. I shot one deer, which we had cut up and frozen for special occasions. I had a nice double barreled shot gun that I had obtained by trading my two pistols for the shot gun. When overseas, pilots were issued the standard 45 Colt. I also bought a small Beretta pistol in Italy, both of which I brought home. But when Barbara and Patsy got to be 5 and 3 years old, I decided to get rid of those pistols, lest those kids find them in our gun hiding place, and I think I made a wise trade.

In May of 1953, Camp of the Woods was sold out for our favorite weeks, so we explored the idea of renting a cabin near Camp. We drove from Buffalo to Speculator in a May snow shower. We met Mrs.Teft, the Camp photographer and friend, who told us about the Meter property on Golf Course Road, along Lake Sacandaga., just 3 miles away. We found a woodsy lot with a big old garage on the roadside and a wooden boathouse on the beach front.   There was also a tall, round stone chimney, left standing alone from a fire in the 1920s. We couldn't get inside the buildings, nor were there any windows. The only pathways led you to the sand beach, or to the two- seater outhouse, because the property was so cluttered with brush. I called Mrs. Meter in Northville. " Oh Pinkie" she said, " Come down and have dinner with us, and we will show slides of the camp". Before we left that night we agreed to buy the whole thing, with 200 feet of water front, and a quality hand pumped water well. We went back to Buffalo, all excited, and secured a larger mortgage on our home to raise cash for the closing. The banks would not grant a new mortgage on just "land". A month later, I was called in to the N.Y. headquarters, and given a promotion, to be the Prestone Division Sales manager of seven states, requiring a move to Atlanta. Oops!

So instead of a July two-week vacation to direct the builder to move the garage and the boathouse down to the chimney, I was working in hot Atlanta. I had eight salesmen and scores of key distributors to get to know. Atlanta was still quite rural in 1953, and real nice homes were affordable. Ours was 1 ½ miles from Atlanta Bible Chapel, associated with the Brethren. There were a lot of couples our age at our new church home. It was a young, family-oriented, evangelical group who really desired to be a witness in the area, and to worship the Lord faithfully. We went to all the Sunday morning and evening meetings as a family. Our Betty was born while we lived there, and in those seven years, our kids grew up in Atlanta schools, where they still had the old-fashioned school mom as teacher. It was a really great time of their lives. We all got to see a lot of the South, and learned to love the southern people even though they still called us "Yankees."

For seven years, Gwen and the kids spent the whole summer at the lake. Our big Ford station wagon was just right for the 1,800-mile trip north, and back. Boards on the back floor, provided the extra bed needed for our four youngsters back there. We usually left Atlanta at about 5 PM, and drove all night until 11 AM the next day. Gwen was a good driver, so we alternated driving every two hours. One year we even trailed the Comet sailboat down to Atlanta so we could put a double layer of fiberglass on it. After it was sailable again, we took it up to Lake Lanier, at John and Mae Brown's cottage. With the mast still up, we the pulled boat and trailer out of the water and drove down their short street, until CRACK: the mast broke in two. How did those overhead street wires get there? I had to send to Boston to get a new mast.

We loved the "Gone with the Wind" atmosphere of Atlanta those days. And we traveled to the country towns, to take in special attractions like Calloway Gardens, FDR's summer home, Stone Mountain, and the charm of Savannah. Also, it was easy to become a Georgia Tech fan during football season, as Tech's games were headline news, and we had 50-yard line season tickets reserved by my Prestone office. The kids had some fine neighborhood friends to play with then. And special holidays like Halloween, was a big neighborhood event in which to enjoy all that good "southern hospitality". They all learned to ride their bicycles, roller skate, and even play touch football with me on our front lawn. Those were some of our best years together as a family. We loved the tall pine trees, the azaleas, and the pretty magnolia trees in our front lawn, too.

But it all came to an end when I was assigned as Prestone Division Manager for the northeast, out of the New York office, and would have to move back "home." At each move it seemed harder to leave our church and our special Christian friends. What a patient, understanding wife and mother Gwen was. She took it in stride, but we enjoyed together the excitement of it all: what would the Lord have for us to do at a new location. My office would be at 270 Park Ave. so we could choose Connecticut, Westchester or New Jersey as places to live. Our first consideration was a good church home for our family. Gwen and the four kids had adjusted well, so far, on each move, but now the kids were growing up and needed a progressive Sunday school and other young Christian friends. We found that at Woodside Chapel in Scotch Plains, N.J., so we signed a contract to build a new house a mile away. Soon after we got there, Patsy, John and Betty were competing on Sunday morning radio Bible quiz programs, which Dr.Gill coached, so that Chapel kids could compete with other church kids on the NBC radio station in NYC. We all grew spiritually in that Assembly, and we attended all the meetings as a family. We appreciated dedicated families like the Shetelichs, the Dicks, the Gills, the Mayers, the Lotts, plus Irving and Carolyn Hansen, Gwen's sister.

It was our first home with new construction, so we ended up with a five- bedroom split level, big enough for Barbara's young peoples' group to have some great dinner parties, as well as great fellowship times with the Lord's people. The biggest event perhaps, was Barbara's marriage to Jack Gill, the prince charming of the day, in1965. What a fine husband he proved to be. Their wedding list required the Hydewood Baptist facilities to handle the 300 guests, plus more than you would ever believe of Gwen's homemade turkey salad for the reception. Gwen and I cried at her wedding, a little bit because she was the first to leave our happy family life, but mostly out of joy for her, as she started a new life with Jack Gill.

NEW CAREER WITH NAPA

But storm clouds appeared soon after. As a result of four years of commuting over three hours per day to New York, plus the sense that there was a better way to make a career, I decided to give up my 28-years association with my fine company, and go out on my own. Not many men were leaving good companies like Union Carbide those days, but I had confidence there was also a future in a business which could be managed and controlled by myself. I had known many NAPA automotive distributors, and the integrity of the NAPA top people. It was the largest automotive warehousing system in the country. They assisted me in buying three NAPA stores and adding a fourth, from an honorable man who wanted to retire.

Gwen was quite apprehensive about leaving the security of the Eveready income for an unknown future of my own venture. But she pitched right in to help on many of the necessary things that she could do at home. I remember how much fun it was to have a first "business" dinner with her, in a company car, on our company credit card. When my son John graduated from college, he managed one of the branches, and did a good job. The Lord seemed to bless the enterprise in many ways, and I found that I had more free time for the Lord's work, more money for the Lord's work, and many more ways to serve Christ in our local church fellowship. So another move.

All our savings had gone into the business when NAPA arranged for the bank loans, and we also had to find a place to live. But we had only $3,000 left to live on. So we went to the Fair Haven real estate people to seek a rental house to tide us over. The agent had no rentals, but took us to a lovely street, and showed us a pretty Cape Cod house that could be purchased for $2,800 down and a $26,000 mortgage(this was 1965). The only catch was that the owner could not close for four months. We signed the contract, found a temporary rental, and just waited out the four months to take title. Patsy, Betty and John were in the Fair Haven schools then . Patsy later graduated from Houghton, had majored in French, and was happily employed by the United Nations. It was while living in New York that she met her Ted, and later they were married in the church where they had met and served. Their choice for a reception dinner was definitely not the church or a restaurant. They found a loft building apartment in which, the owner, an artist, rented out for such occasions.

Fortunately, my business truck could fit all the chairs, plates, food and utensils, needed for the reception. Patsy and Ted bought all kinds of breads, cakes, and special foreign dishes. The cooked lobster, 13 pounds of freshly cooked meat, was prepared by the wife of a lobster fisherman we knew. It was fun to walk a few blocks from their large church on South Broadway to the reception, then take the loft building freight elevator to where the sounds of Robin's flute, was playing quietly over the din of the mingled guests. The unusual tastes, and the unlimited varieties of the best of New York delights, made it a reception to remember. Patsy and Ted continued to their careers in New York, Patsy, as an editor at the United Nations, and Ted as an interior designer for one of the large architectural firms specializing in office buildings. Eventually, Ted returned to his first love, college-level teaching, where he excelled at Hood College, University of Oklahoma and Oklahoma State University. Not only was he voted as the best dressed professor by his students, but he has been highly recognized for his presentations and papers given to the American Association of Interior Designers, his peers.

Our new home on Buena Vista Ave. was bounded on each end by two tidal salt water rivers. John sailed our Comet there, in the spring and fall , and in the winter when the ice was more than 6 inches thick , we sailed our ice boat. The oldest ice boat club in the U.S was in Red Bank, enabling members to sail from the ice to a comfortable club house, and share in the fun of moving at four times the speed of wind(at 30 to 50 mph). One year the ice was so good, and the weather ahead so promising, the Ice Boat Club declared a National Regatta. This brought ice boats out of the woodwork, some coming from Michigan, Maine, and even Europe. Huge 50-foot long mahogany beams, one foot square, on three steel runners, comprised the ice boats from the Poughkeepsie Hudson river clubs. They sailed gracefully with four people aboard. It took four men to push the craft, even with full sail, fast enough to gain the momentum needed for the big gaff rigged sail to pull it at cruising speeds.

John built a dandy racing boat from parts which a retired ice boater had scrounged from members. It was a two-seater. John and I painted it all up and assembled the beautiful sail, mast, runners and beams that comprised it. John sailed it a number of times that year, sometimes flying, with one runner up in the air. But as the winter faded, the ice got a little thinner from the tides underneath and the sun. But we wanted one last run together. So we sailed out of the Club harbor, along the shoreline, until " CRACK", the runners went through the ice, and the boat started to sink. John jumped off, wisely, on the side where the ice looked thickest, and stood up, while I went deeper in the water until the runners submerged. He yelled, "get spread eagle on the other side", which proved to be better in order to work my belly on top of the breaking ice, until it would break no more, and I could stand up. We decided to leave the boat, sail and all, sinking in the water. As we walked back to the clubhouse, dripping wet, we heard first- aid sirens coming to help. They checked me over for hypothermia, and sent me home for dry clothes. That ended John's nice iceboat, as when the ice melted we retrieved it in quite a sad condition, so we gave back to the ice boat club the pretty sail, mast and runners to the next adventurers to rebuild once again.

Our new church home became Fifth Ave. Chapel in Belmar, 30 minutes away. That was a blessing, for we found a friendly group of the Lord's people there, anxious to spread the Gospel. It wasn't long before we were walking up and down the streets, placing tracts or invitations on porches. Even participating in outdoor meetings on the sand swept ocean boardwalk, one block from the beach. There was a time when fine a British evangelist was invited to preach every night at our Chapel for two weeks. Families brought their children, and their neighbors, so that many unsaved folks came out. Some 25 people came forward for salvation, or for assurance counseling on their faith. Dick Saunders, the evangelist, reaped where the seed had been sown in many lives, and God was at work. My assignment was to train the men and women counselors, and see that there was proper follow up with those that made decisions. It was great to be busy for the Lord, and I enjoyed serving as Sunday School Superintendent for several years, too.

About this time, Betty went off to Barrington College in Rhode Island, were she spent a year and a half, until transferring to Gordon College, where she took her degree. Those summers she served as counselor at Camp of the Woods, and made some lifelong friends. Betty was the youngest, always enjoyed travel. Who else got trips to France, Austria, Poland, Russia, Sweden and Finland for college credit? And long before mom and dad could go there themselves. But though her training was in history, she excelled in administrative and accounting positions in the medical world. Now that she has some of dad's old tools, there is nothing she can't fix. Or at least try, although, sometimes it means a call. "Hey dad, what do I do now?" Which is always okay by me, as what are dads good for, anyway? She loves big mortgages, big houses, big cars and big vacations and works hard to get them all, giving to others in her own special ways.

A new Brethren chapel started in Red Bank, so we felt it advisable to support that work so close to our home. The Belmar elders were in agreement with that and we continued in happy fellowship with them over the next 25 years. The ministry at Bethel Bible Chapel was struggling in the early days, and we wanted to be a part of it . Gwen helped by playing the piano then, and later took organ lessons to fill in for Bernard Gwilliams. One day, John called me from Taylor University, where he was in the middle of his sophomore year and doing well. He said "Hey dad, I've met a nice girl named Joyce Kegg, and we have been talking about getting married, what do you think" I coughed and choked while I got composed to "think", and said, " John, I know you are mature enough to know the right girl with whom you want to share your life with and I pray the Lord will guide you both." The next August they were married at Joyce's church in Kennett Square, Pa. the mushroom capital. They have been serving the Lord, faithfully ever since. The Chapel ministries were reaching the neighborhood, in Red Bank, until a devastating fire destroyed the whole building. We thought it may have ignited by vandals. Many brethren Chapels sent us generous love offerings, neighbors sent checks and our insurance enabled us to rebuild. Everyone pitched in to help. Pat Truglia and Ernie Fox encouraged us all "to do it unto the Lord". I chaired the building committee, and spent a lot of time with Architects, builders, and town committees to complete the project.. Meanwhile, the Firemen who put out the fire, offered the basement of their firehouse one block away, for us to continue our services.. That summer we still had a DVBS on the property, by renting party tents for the kids and leaders to conduct normal activities and meetings. It was encouraging when two young sisters from the neighborhood got saved, They both told their parents that they needed the Lord also.. The parents soon came out Sunday mornings and trusted in Christ as Savior. The Horvaths grew rapidly in the Word and were a great help to the Lords work, until the Lord took him prematurely.

It seemed our church was growing spiritually, but slowly numerically, so several Elders prayed about it. We prayed that a gifted young man could be found, to assist in the preaching, teaching and other aspects of our ministry. Not long after this, a fine young fellow, Bob Billings, showed up at our Wed. night prayer meeting. He said he was a student at Gordon Conwell Seminary, on a break, and lived nearby with his parents. After graduation, Bob taught in a Christian High School in Detroit, and also preached at a number of Brethren chapels in Michigan. When he desired to come back to our area, we encouraged him to assist in the teaching and preaching ministry of our Chapel which he does effectively. We have been blessed by his ministry and friendship. During our later years, Gwen played the organ, working closely with Betsy Pierce, on the piano. What an asset Betsy also has been to the Chapel ministry, as she provided quality music, plus her personal charm in her many other involvements. Other faithful members like Jack Malonson and his wife Millie, Ken and Cathy Kuppler, Pat Truglia, Micky, and Rick Ings, contributed unselfishly to the ministry at Bethel, we thank the Lord for them often.

LAKE PLEASANT N.Y.

Our 2nd home at Lake Pleasant, provided our children with opportunities to expand their lives in so many ways. Not only was it a time for family fun and bonding, but it was a place where many of their Christian values were gained by our being together. These same values came from involvement at Camp of the Woods, attending Tapawingo girl's camp, or Deerfoot Lodge boy's camp. Since 1953, Gwen and the 4 kids arrived at the end of June, and stayed until Labor Day, while dad could only make it for the usual 2 week vacation, and a few weekend trips up there. Gwen watched over them, gently guided them, and made the Lord's presence in our home, be a way of life to them. The water, the beach, the boats, the daily Chapel services, the Camps, the cook outs, were all theirs to enjoy. Camp of the Woods had a season membership arrangement, so we could use the facilities, the beach, the concerts, programs etc. We often met family or friends who were vacationing there.

One summer, Gwen's family had a reunion at Camp, so we arranged a corn roast. Our son John brought up a couple of sacks of the best corn on the cob any one ever tasted. It was picked that morning in NJ. Then boiled over an open wood fire, in two huge pots. Mom topped off it off with some hot dogs and sauerkraut. Makes a pretty fine evening. Over the years we often cooked outside by the waters edge. Sometimes it was lean medium rare burgers, for a family get together. But on one occasion, about 40 friends from homes around Lake Pleasant, came for a Lobster party. In the days when Lobsters were plentiful, one of our customers was a lobster fisherman who went off the Jersey coast with his lobster boats. 75 lobsters, and a case of steamer clams was ordered from them. We kept the lobsters and clams alive with plenty of packed ice, while driving upstate, the day of our party. It took 8 oak logs and 3 large pots of boiling water to cook those red lobsters the required 5 minutes. About 40 hungry lobster lovers put them away fast, but no one even tried the clams. When we finished the desserts, everyone gathered chairs by the campfire so we all could sing our favorite hymns or other songs, while watching the sun set across the lake behind the mountains. Some of the fellows sang "barbershop" quartet numbers, and that made the evening. Sometimes, we told stories of recent happenings, or other startling events, while the bonfire crackled away.

Over the years, there were many occasions like that, (without the lobsters) and most times it was just sitting around the roaring fire, until the sun set and the fire burned down to glowing coals, that were just right for Mom's (and everyone else) favorite "s'mores" If it was just us Pinkhams, we stayed until the cold night air flowed from the woods, and it was dark enough to watch the planets in their orbits, the milky way overhead, and the always dependable north star, steering our vision to Shuttleworth's place and beyond, to an occasional dancing the Aurora Borealis.

In the winter months, the Adirondacks used to be a place nobody wanted to be, except the 500 or so local citizens who made a living in their professions and in winter activities like cutting ice blocks from the lake for storage, ice fishing, deer hunting, trapping, and snow clearing. But very few people with summer "camps", ever intended to come up to Lake Pleasant's deep snow and sub-zero temperatures. Dr. Armstrong, Sr., Dr. Ed Armstrong's dad, was the only one we knew in the 1950s. Many people with a summer homes, had built their place on cement peers, most of which, moved up and down, or were tilted by the ice, if they were not dug more than 4 feet deep. So that's how we constructed ours. We had cement peers built, so there was a 2- or 3- foot crawl space underneath the house. We didn't even consider putting in a furnace, as the big stone chimney fireplace would be enough. I helped the builder dig many of the peers, and got so tired one day I had to quit and go for a swim with the kids. When I came back to dig again, I said to the Chet Rhudes, " after this what are you going to do tonight"? "Oh," he said 'I'm going roller skating." That same year he was working for me, he was full of anxiety, because there was huge forest fire in the Moose Lake area, which if it got out of control, would force him to leave his work, and volunteer his time to put out the blaze.

For the first 25 years we seldom drove up in winter. But when we did it was awesome to see the snow 3-feet deep all over the property. The snow plow had piled up a wall on the street so high we could hardly get over it. We bought several pairs of gut caned snowshoes, which kept us from falling through up to our thighs. Cross country skis were sometimes better than snow shoes for getting around. The weekend we stayed in the old Osborne Hotel was fun for the kids, as they brought their ice skates and had a great time skating on the big rink on the lake. The hotel had plowed the snow on the ice and watered the surface for smoothness. The hotel, now gone, was then a landmark three-story building, jutting out on the lake by the Kunjamuck River bridge. Gene Tunney had trained there in the twenties, over a number of years. Even in the 1930s, other fighters like Schmeiling, and Maxy Bear stayed there. In 1935, we could just walk across the road from Camp's entrance to watch Schmeiling spar with a big black man. Then they did their running off through the paths in the woods and logging roads, training for the upcoming Schmeiling – Louis fight, when Joe Louis was in his prime.

As more people came to Camp of the Woods, more folks wanted to buy property, or buy a choice house, so they could be near Camp to take in the splendid concerts etc., plus the spiritual value of the Chapel services. Over the years so many ex-campers were homeowners, Gordon Purdy had to set up fees, rules and agreements so the guests would not suffer from overcrowding at certain places like tennis courts and concerts. So the "associate" members would now have to pay their share. As a result of the high demand for properties created by ex- guests, real estate values have risen considerably more than other beautiful lakes in the Adirondacks. In about the 1970s, Camp decided to winterize, by building a new office, bedrooms, dining room, swimming pool, which Wes Nilsen designed with Bob Hansen's engineering help. It was then called Woodlands.

Since we still did not have heat in our place, we made a reservation for the first New Year's weekend at Camp, the first year Woodlands was built. The snow was deep, the ice was thick for ice skating and the speaker was excellent. However, at 2 AM, the room next to us had a huge leak in the hot water pipes under the window. The occupants had gone to sleep with the window cracked open a bit. The "10 below" cold air dropped on the pipes and the freeze split them. Gordon and others were in a commotion, but fortunately one of the guests, a plumber, had all his equipment with him, and saved the night for us all. That weekend gave us a good taste for the snow up there.

The ski mountain, 10 minutes away in town, was not only easy to get to, but quite affordable at $20, compared to the popular Gore Mountain 45 minutes away, and $50 a day. Snowmobiles were getting popular, so we decided to winterize our place to enjoy riding in the snow. The only place we could place a furnace in our two-story place, was to put it upstairs where it was un finished, and build a second bathroom there at the same time. The hot air was forced downstairs to the crawl space where it was distributed to every room. A Norwegian wood-burning stove was placed on the hearth in front of the fireplace, so the smoke pipe could easily angle up the chimney. It put out so much heat, the furnace hardly came on when it was just above freezing outside. We had to buy a cord of split hardwood like beech, birch or oak for our Yotel stove. We even cooked steaks, mickies, and onions over the red coals any time we liked.

Soon Barbara and Jack, Patsy and Ted, John and Joyce, Betty and Gwen, all had to get skis and ski boots, as well as cross country shoes, helmets, face masks and gloves, and find a place to put them all. The grandchildren had more 'stuff' too. Eventually, snowmobiles were "needed, as the lake was a highway for the snowmobiles, as long as there was at least six inches of ice all across the lake.(we determined that by waiting for other snowmobiles to be out there first). The State of New York made wonderful trails from openings off the lakeside to all kinds of places, by groomed trails. They groomed the trails to many directions, after every snow storm.

The snowmobiles were also useful to go out on the lake and check on the fishing success of some of the ice fishermen. I got to know an old timer resident, Louis LaVornway, who sometimes wanted to get his fishing shed on our Lake through my property. His snowmobile could pull the shed (with skis under it) to any of his favorite locations on the lake. He lighted a fire in his homemade wood stove, and then set out to drill five holes in the ice, in a circle 20 feet from the shed. That is the New York State limit on fishing holes. He put live minnows on the lines of each tip up, inserted the reel in the hole, pulled the latch to a set position, and just waited for the fish to bite. One time I went out to see how he was doing, and he showed me a nice 28-inch wall eyed pike laying there on the ice. "Come on out and fish," he said.

So I drove my snowmobile back to the house, ran in and got five tip ups, and joined Louis in the warmth of his comfortable little shed , just enough room for us two. I started to drill the ice holes with my nice 5-inch hand Augur, but Louis said he could do the job quicker with his gasoline motored augur. So he did, and we dropped the live baited tip ups in the holes. Then went back in the shed to warm up again. Only five minutes had passed when we spotted one of my flags in the up position, indicating a bite. I wanted to rush out there , but Louis said "slow down , let the fish swallow the live minnow all the way down". After about four minutes, Louis let me give a slight tug on the line to determine if the fish was still on the hook. Sure enough, it was. So, after cleaning the ice chunks on the hole with a spoon, it was time to reel in the prize. And it sure was, a nice plump 18-inch Rainbow trout. Louis threw my fish over by the rest of them on the ice, but I said "Louis isn't it a good idea to gut the fish first?" He shook his head, so I cut the belly open, and there was the live minnow the fish had swallowed. When I started to throw it away, he grabbed it, threw it in his minnow pail where it happily swam vigorously with the others, while Louis said "we'll use that baby again"!!

CAMP OF THE WOODS

Gordon Purdy was a young, first-year counselor in the summer of 1933. And we both served under Pop Tibbitts the founder, with fear and trembling, for another few years. Gordon eventually became Pop's secretary, went with him to Park of the Palms in Florida, and in Pop's old age became his right-hand man. In 1946, when Pop died, only Mrs.Tibbitts, and Miss Emma, were on the board of Camp of the Woods. So they appointed Gordon as chairman and CEO of the whole operation. And he got the "keys" to the place. Because Gordon had been a new Christian only a few years, it was a daunting challenge to him. Dick Woike, the owner of the island on Lake Pleasant, and a good Baptist, was a faithful mentor to Gordon those early Camp seasons after Pop died. In the winter months at Florida, Fred Sacher, a mature Christian (Brethren) was a spiritual mentor to him. Gordon's early years were such a challenge to him, he never forgot them. He kept in touch with many counselors of those days, watched their careers, accomplishments, and reveled in his association with them. He was genuinely proud of his friends of the 1930s. Often he would call me up and say, "Jack, I just hung up from talking to Bob Rosevear; he's conductor of the Toronto Symphony now. Come on over and meet him." Another time it was Dr.Ev Fuller of HCJB, Phil Reid, or Crow Stahl, people whom he admired. And the older he got, the more nostalgic he became about the people and events of those early days. He loved to tell stories about them too. His favorite story was about Gwen's family, all nine of whom spent a week's vacation at Camp in 1933. They were with a large number of families from the Yonkers Chapel, about 30 in all On their last night at Camp, they planned a party on the white sand beach, of the barren little island, a 1¼ mile canoe paddle away. The butcher in town ground the hamburger meat, and even asked if he could "add a little pork in it for flavor". They cooked the burgers over an open wood fire, had a great time and paddled back in the moonlight, to their tents on the Camp beach. Within a few hours a few young people like Gwen, were sick to their stomachs, but some of the parents did not get sick until on their way home the next day. Gwen's mom and several others went in the hospital with severe symptoms, cramps and fever; others were suffering various degrees of severe illness, which gradually diminished. The diagnosis was "trichinosis" from the rare pork in the hamburgers. Fortunately, they all recovered, and Camp food was not to blame, as Gordon would proudly point out.

Those early days in the thirties were special to me as well, as I remember the Dunkerton family that year in 1933, because Gertrude, Mom and I were there the same week. The Dunkerton twins, June and Joanne were the talk of Camp and were cute as could be, although I never noticed Gwen then.

Camp was primitive in those days. When a guest arrived, he was escorted to his tent by a counselor, who carried bags in a wheelbarrow. In the tent, a candle, a white washing bowl, a slop pail and a water pail, and a path to the toilets, was standard fare. The tents on the beach were reached on a long weather-beaten wood walkway. The tent floor was the same level, and the tent sides were wide boards, about three feet high, rigged so the canvas roof and sides came over a wood frame to the sidewalls. The first-year counselors were in charge of sections called "tribes" and had to make the beds with fresh white sheets, and keep guests happy. That included rushing to the tents every time a big storm whipped up. Tie the fly down, tighten up all the loose ropes to hold the flying canvas from blowing away.

One Saturday when the guests were arriving, I was assigned to a goony-looking guy that to a 16-year- old kid, looked like Frankenstein. I had to wheelbarrow his huge bags up the hill to a room down a hallway that I was scared to go in. I hastily accepted his tip of a quarter, and ran out of there. But for my four years there, I loved every day. Played reveille after the 7 AM bell, played a bugle "church" call for morning chapel, and taps at 10 each evening, except Sunday, when our trumpet quartet played "Now the day is over", from the office beach. The constant challenge was to please Pop Tibbitts, the director. Every evening meal, four trumpets, and four violins lined up by the dining room fireplace, waiting for Pop to hit the chimes, signaling us to play the music through once so the whole 400 dining hall guests could sing the words of "God is great and God is good, And we thank him for this food . By His hand must all be fed, we give thee thanks, for our daily bread."

I could barely keep up with some of the good trumpet players, like Crow Stahl. He was a music major at Eastman School of Music. Occasionally, he and I went out in the woods to practice duets. The acoustics in the trees were awesome, because the notes actually made harmonics sounds, somewhat like octaves in your ears. Crow became the band director for several years, so he had to provide the band music for the outdoor Saturday evening programs. The guests enjoyed sitting on blankets in front of the band platform up in boy counselors' park. The program was light, brightened by a huge log Campfire, and an amateur gymnast dressed in white pants, doing kips on the cross bar, while the band played "Stars and Stripes". But Camp wasn't dependant on professional talents, we came for simple needs, like love of the great white sand beach, the talents of the average church kid who would sing for the first time to big audience, and was so terrific, you hoped that kid would be back next year. We also came for the underlying reason that Camp existed, expressed in the camp motto "Vacation on Purpose". Chapel service each morning started the with great music and the Word of God. All the guests attended and all the counselors were there on the platform, too.

So as Gordon developed Camp he sought additional Christian men to be on the Board of Directors. In my 30 years on the board, some of those years were as vice president and some as secretary. Since offerings and gifts were $100,000 each year, most of it was distributed to board-approved missionaries. But since $30,000 went to a Hong Kong children's camp, Gordon and Anne made a trip there almost every year. Gwen and I went out there three times. One of the trips, Gordon Purdy arranged a three-week route by way of New Zealand (1st week), Australia, Bali, Jakarta, Singapore(2nd week) and the Hong Kong camp, Tokyo, and Hawaii (3rd week). Literally, thousands of young people have heard the Gospel at Suen Douh camp, and a high percentage made decisions for Christ. To this day, it is still operating as it did under British rule, in spite of China's uncertain rule.

Camp of the Woods was directed extremely well by Gordon Purdy for over 50 years; he never owned it, as it was a " not for profit" organization and over the years was controlled by the interdenominational board of some of the finest dedicated Christian men: Walt Maloon, President, Correct Craft; Wes Khurt, President, Sikorsky; Phil Bauer, President, Tastycake, Ralph Larsen, President, J&J; Norm Sonju, General Manager, Dallas Mavericks, and John Kubach, a Wall St. banker, all of whom are the "who's who" of America. There was another non-voting board, which also was composed of very fine men who met only once a year in an advisory capacity. All of these men brought their families to Camp and some like Ralph Larsen met his wife there. Another good friend of Camp, John Maltese, was married there a few years ago, to his lovely wife, Anna. John was Camp's favorite violinist, and still is.

My best man at our wedding, Ted Carlson, didn't get married there, but he spent his honeymoon at Camp. But one night they got lost in the fog in a canoe, paddled in circles from the island until 6:00 AM. There were other great supporters of Camp, too. Like Wes Nilsen, Joanne's husband, who drew the architectural drawings for the big dining room, swimming pool, office, and deluxe rooms of the Purdy center. All for pennies in terms of expense. His friend Bob Hansen worked out the engineering factors and put the seal on them too. Then another good Norwegian friend of Camp, John Johansen, wrote a beautifully written book on the history of Camp. All of these are a small sample of the fine people that have been "campers" over the years.

There were great preachers too. It would make another huge list of "who's who" in the " preacher " world, as the finest and most sought after men in the country have come to preach from God's word, a relevant message at the morning chapel service. Just to hear the messages of a certain speaker, many guests have returned year after year. Of course, others have come back because of the great music of the college students led by the Winklers or the talented performances of Jim and Carol Rumsey, or the fascinating forums by Betty and Wayne Frair.

One can not mention Speculator, N.Y., without also referring to Deerfoot Lodge, a Christian camp for boys, which has been judged by many, as the finest boys camp in the country. Deerfoot was founded by Alfred Kunz in 1932, with the financial support of Harvey Wadham. Deerfoot had a difficult beginning, because it was in a remote location, and had started during the big Depression. A two-week camp experience, led by competent staff with a spiritual emphasis, has developed thousands of "men of the master" over the years.

In the 1970s, Jack Gill served as director for five years. Though he was a high school science teacher, he made significant strides in those years. One of them was to recommend that a permanent director be hired. Jack and our daughter Barbara worked year round on Deerfoot matters, which made their home and their time occupied with recruiting 50 to 70 staff, purchasing boats and docks, follow up and registration mail, and a host of Camp issues. Jack was a boy's man, who inspired young men to achieve stature in outdoor sports, and spiritual discernment.

When Governor Pataki spoke in Speculator in 2007, he said "I love this part of the Adirondacks, and I have sent my boy to Deerfoot, and my daughter to Tapawingo." Chuck Geiser later admirably directed Deerfoot for the next 20 years. Jack dedicated his time with his family, his church as an elder. And to improve his teaching, he worked on getting Masters degrees in meteorology and oceanography. He also took a course in creationism so he could tell his students there was an alternative to the theory of evolution. He was careful in teaching both aspects, and did so for several years until another teacher complained to the principal. When the principal reviewed the issue with Jack, he said, "Just keep on teaching what you are teaching." Jack was on a beach vacation when at 53, he collapsed and went suddenly to be with his Lord and Savior. Wendy, Cindy and Brian with his Julie, have been an enormous strength and comfort to Barb ever since. Evidence of Jack's legacy is seen every day. Deerfoot Lodge constructed a huge log cabin "lean to" in his memory, out on the point, for boys to use.

RETIREMENT WITH A PURPOSE

While our home church was our main focus of service, there also were other opportunities in which to serve the Lord. I had set a goal of 65 as my age for retirement from my business.

I had known too many professional men who did not know how to retire. That is, they liked their work so much that they could not give it up. Gwen and I liked to travel, but we also wanted to be involved in the Lord's work even if it were part time. So we began to seek a mission type experience. Jim Gillette in the Dublin, Ireland, area came to our church and pointed out all the needs for prayer, and for people to come and help. We went for two months the first year, then we went twice the second and third years, then once a couple of years, for a total of seven years. We traveled in Europe, Great Britain, and Ireland while we en route to and from our mission station.

The ministry, called Ireland Outreach, is located 7 miles south of Dublin, at Dalkey, near the Irish Sea. They are involved in evangelism, distributing literature and Bibles etc. to the schools across the country. The public schools were very poor economically, mostly taught by priests and nuns. They were required to teach "religion" three hours a week, but they had very few helps to teach from. So Jim Gillette had the vision to provide Emmaus Bible lessons free, and later New Testaments also free. There was a Christian bookstore in the building. It was frequently visited by nuns, who were sent over by the Mother Superior. Their Catholic center was a retreat for nuns from all over the world, and was only a block away on the pretty blue water side. When we arrived the first year, the headquarters building was filled with part-time workers like us, so Jim took us to an abandoned villa, which a friend of Jim's had given permission to use. Jim helped us up the high steps to the chilly rooms inside. He showed us how to light the propane burner, as there was no furnace. It was cold, too. Then Jim gave us a little lecture on the gypsies. He said, "Don't let them inside if they knock on the door. Don't give them anything either. They case an area and are prone to steal." With that good news, he left us. About 10 minutes later we caught a brief view of a man walking around the house and who then disappeared. "Who's that?" That's when we said, " my goodness, the gypsies are already here". And we wondered what were we doing there? In another few minutes, there was a knock on the door; the man on the outside said he roomed downstairs and wanted to get acquainted. What a relief! I told him how glad we were that he had let us know where he lived. When he left, we opened up our Bibles again, and read Ephesians 2:10, and that night had the most restful sleep under the moon light and "under his wings".

We performed most duties over at "Charleville," the large headquarters building. It was a castle-like granite-walled structure three stories tall, with large, attractive arched windows. Inside, there were huge conference rooms and areas to partition into bedrooms. All the part-time mission workers ate there, three hearty meals each day. The cooks were volunteers from the States or Great Britain. Actually for three of the years we came , Russ and Mary Lou Farwell, from the Kansas City area, cooked for about 20 hungry folks, along with other duties they fitted in. Gwen was assigned to work on the Emmaus Bible courses, grading lessons and answering responses from all over Ireland and even from other parts of the world. Since so many were Bible courses given to schools, some of the courses were sent back by the nun teachers, in a batch, for the whole class. Others came from places like Nigeria, as a result of a visiting nun taking the Emmaus courses back to her country. This occurred quite often. The Mother Superior was very supportive of the I.O. ministry and would send nuns over to our Christian bookstore for all kinds of Christian books and supplies. They welcomed us there ,too. Although, they may not have known about some of the other evangelistic efforts, such as Dublin street corner meetings, hotel evangelism using Christian films and personal witnessing that was going on. (Catholic families would come to a hotel for such a meeting, but not to your church).

About the third year we were at Ireland Outreach, the building inspectors discovered what they call dry rot, in the beams and joists of the majestic Charleville, the headquarters. A moss-like growth gets on the under floor beams and window frames on most all 100-year old cathedrals and buildings in Ireland and the British Isles. This meant the whole three floors, windows and all, had to be gutted down to just the two-foot wide granite walls. The roof was not affected. An Irish contractor did all the demolition and rough reconstruction, leaving all the plumbing, electrical and wall board finishing to be done by volunteers. It was amazing to see how many fine capable workers came to help. Most came from U.S. churches, some from Canada and a few from Great Britain and Northern Ireland. Since I had some experience in taping wallboard seams, I stuck with that work for a couple of years until it was finished. Then came the painting. I preferred to work alone, plastering, etc., but since Jim Gillette would let almost anyone help, including some young people, supervision was necessary. He gave that part to me too, which slowed me down. One two-month spring session, I had taped a few rooms, but there were a lot more to be done. When Gwen and I returned in the fall, I noticed some strange tape on a few large rooms where the tape was falling out, not sticking. What a mess was done by the unsupervised young people who used some wide electrical tape instead of the proper paper tape. I had to rip it all out. It was fascinating to see the building take its own character again.

A spiral cement stairway had to be built, so Jim got a local named Paddy Purdy to make the wooden frame for the cement structure. Paddy did a great job framing the spiraled steps from the basement to the third floor. He worked alone, but he had a tendency to drinking heavily, but never on the job. He would work Monday and Tuesday, say "good night, I'll see you tomorrow", but tomorrow never came, until Friday. He would come back to work, with no excuses, just like nothing happened. When he completed the spiral framing, all the volunteers started carrying buckets of Paddy's cement. They went up a ladder to a window, where the buckets would be unloaded in the spiral stair forms, so Paddy could rake it in and smooth it to a fine finish until it completely dried.

Then president of International Bible Society came through Dublin to visit on his way home from a conference. He was impressed with the freedom which Ireland Outreach had to get the scripture in the hands of Irish families, by way of the schools. The outcome was that IBS agreed to ship 100,000 NIV New Testaments a year, free, to Ireland Outreach. Praise the Lord, now I.O. could give away to any school, the New Testaments, plus Emmaus Bible courses. It was exciting to be there when a tractor trailer pulled up with 50,000 items to be unloaded. And this continued until 500,000 New Testaments were shipped 4 years later. But it was even more exciting to see a huge van with letters printed on it "Dublin High School", drive up to pick up nine cases of New Testaments.

There was always a need for workers to handle the many tasks of I.O. One of the good ones was a 17-year-old boy named Flan, short for Robert Flanagan. His father, in one of his many drunken rages, had put him out of the house and not for any serious wrongdoing. Flan rode around on his bicycle a few days, and eventually came to Jim Gillette because he had been to a Gospel effort in the area. Jim offered Flan room and board without cost. In God's time, Flan trusted Christ as Savior, and stayed around to help in the ministry. One of the periods of plastering and painting, Flan and I worked together. In mid morning and mid afternoon, we always took a break for tea and a couple of Oreo-like cookies. Those breaks provided good times of mentoring, twice a day. Flan was full of questions about the Bible, and of life. So, to get God's view of things, we both decided to read the book of Proverbs on the tea breaks. A chapter a day, in 31 days; we both received a rich blessing. Flan continued being faithful to the Lord, married a fine Christian Irish girl, and occasionally we hear good reports about him.

Through John Kuback, I was invited on other Christian Boards, such as the American Tract Society and the New York Bible Society. I served on the American Tract board for five years, until they moved from New Jersey to Dallas, and it was wonderful to see how God used tracts to win men and women to Christ. American Tract Society has been printing tracts since 1805. In 1864, the chairman was also the commandant of West Point and ordered that a Bible be given to every plebe. This tradition still is in effect, and Gwen and I witnessed the presentation of Bibles, at the cadets' Sunday service at the West Point Chapel on the beautiful Hudson River.

I continued on the board of the New York Bible Society, a very progressive ministry for New York City. The board members were all chosen from men who worked in the New York City area, and were committed Christians. None were chosen because of financial assets; they simply wanted dedicated laymen to serve. The American Bible Society was also in New York and was much larger,, as their donors were mostly from rich main-line denominations across the country. They covered the world with Bibles well, with either the King James or their own Contemporary Today's version. The New York Bible Society was the oldest, however, and served New York City by providing Bibles to all the hotels in Manhattan. Providing hotel Bibles is usually the work of the Gideons, throughout U.S.A, but the large number of hotels in N.Y. required an organization like New York Bible Society to do it, year after year. We actually had representatives paid full time, to service not only the hotels, but also prisons, and foreign ships that had lonely crews in port. They also put Scripture verses in subway cars, using the space where advertising usually occupies. Also, special salaried people serviced the many New York hospitals with Bibles.

THE NEW INTERNATIONAL VERSION – LAUNCH

In the 1960s there was a need for a good more readable text than the King James version. A dozen scholars came to our President Yongve Kindberg, seeking financial backing for a huge project involving over 100 scholars from diverse denominational backgrounds. They needed financial backing, and had been turned down by others but they wanted us to consider underwriting a new contemporary version. The board agreed to finance the work with the small capital reserves on hand. When I came on the board in 1972, over 100 scholars, seminary professors, Christian college presidents, and specialists on certain books of Bible, were already researching the available early manuscripts, plus Greek and Masoretic texts, all on their vacation days or weeks. The final responsibility for all this rested on the top 15 men, a group of self-governing scholars, who were known as the Committee on Bible Translation. They set up review committees to edit the work of every book of the Bible that had been translated, to language standards of the NIV. And there were committees over them, before the top Committee on Bible Translation finally approved the text for publication.

We started to run out of funds, even though the scholars presented us with minimum expenses. We needed millions. To raise funds, we had events like annual large banquets at the American Hotel on Broadway. They became a sort of social event of the year in the New York-New Jersey area as each board member had to invite 10 friends who would be good donors. We did reasonably well that way, but some "friends " gave little, and we had to cover high meal costs and honorariums for the likes of guest speakers as Chuck Colson. Toward the end of the project, the scholars on review committees were very busy (on holidays and weekends). So we approved expenses for such committees to take their wives and children to places in Europe, like Heidelberg and France and Spain, so their work would have privacy and concentration, while at the same time their families were occupied. We allowed this for several summers and it worked well. After 20 years of work on the NIV had passed, we wanted a speed it up, so a little prodding was necessary, recognizing that a translation scholar's work is never done. Finally, the entire NIV was printed. The Committee on Bible Translation did not want to own the finished text, although they would continue on as a committee to correct or adjust it, to maintain the integrity of the NIV. They wanted the New York Bible Society to own the copyright. So when we legally obtained the copyright, we chose Zondervan in Grand Rapids to print and distribute the text. We contracted with Zondervan, giving them exclusive rights to sell the NIVs to bookstores in the United States, whereby we would get a small royalty. The same was given to Hodder &Stoughton in London for the rest of the world. We reserved the right to print and sell the NIV to churches and individuals, providing they gave them away. By selling direct this way, our staff could produce Bibles at a low cost, by the millions to individuals, camps, prisons, military and para-legal groups like Youth for Christ, Intervarsity, Promise Keepers, etc. We printed them for sportsmen in basketball, hockey, football, etc., all with a special cover, and all with helps to find Christ through faith. We soon became known as the "evangelical" Bible Society.

Our small headquarters, at 48th St., near 5th Ave., became inadequate for the increased volume generated by the remarkable success of the NIV. The building was put up for sale at a time when the New York City real estate market was terribly depressed. In the early 1980s, we could only get $400,000 for our lovely four-story structure. A Swedish Society, bought it, and allowed us to continue using the third-floor conference room for our board meetings. Two years later John Kubach spoke with them and in his joking manner said, " We would like to buy the building back." They said " Oh no, the value has gone up, and also the bank next door had recently walked in to offer a million dollars to them if a contract was signed, giving the bank "air rights" over the top of their building. I knew the building well, as on one occasion, I was in the washroom when the last group left the third floor by elevator, and headed for the street. I could not get the old elevator to return and the stairway door was locked.   So my only resource was the windows of the board room. Yelling down to the passing pedestrians from the third floor didn't attract anyone's attention, as nothing really would stop a New Yorker on his way home at 7 o'clock. But I just hung there at the window ledge watching the people go by, when suddenly I saw someone run out of the front door. It was Tony, the elevator man, and when he heard my shout, he returned to let me out. When a suitable building was found in East Brunswick, N.J., our corporate name was changed to New York International Bible Society, in order to maintain the New York identity, but also show that we were now worldwide with the NIV and future foreign translations.

Very soon, our New Jersey location was again too small to handle the volume the NIV was generating, so our board approved moving to a new location. But where? I had retired, so was willing to travel on the search of cities. Several of us went to Grand Rapids, and to Lancaster, Pa., looking for cities with a goodly number of Christians living there to support our personnel needs of about 100. Then I was sent to Colorado Springs with a staff member. We visited other Christian organizations, like Navigators, Compassion, and Christian Booksellers. Actually, there were only eight then (now over 308) and the City Chamber people welcomed us with open arms. I returned to the board with such a glowing report, they wouldn't believe it to be true. Actually, one member resigned because we eliminated Grand Rapids as a location (he had a brother there). Well, when the dust settled, Colorado Springs, though far out west to us New Yorkers, was the place of the future, and the board approved Colorado Springs and a budget for a new building. Wheaton, Illinois, was starting to lose as capital for Christian organizations that needed to network, and soon many others moved out to Colorado. By now, the IBS President and Executive Director and had built a fine, loyal Christian organization. He was aggressive, pushing the board to the edge sometimes, but an outstanding organizer. Always planning something big. Even the $2,500,000 building was never big enough for him.

About 1990, the staff arranged with the Soviet Government to permit us to give 4 million Russian language Bibles, free to the Russian people. The Government under Gorbachev insisted that

2 million of them be printed in Russia. Therefore, in 1991, Stan Black and I, as board members (with Marge and Gwen) accompanied a party to Moscow. Photographers, staff, and friends were needed to dedicate and arrange for the distribution of those precious Bibles. Since no Bibles had been allowed during the 70 years of communist rule, we had to print the old archaic Russian text, as the Government and the dominant Orthodox Church wanted.

We flew all night and arrived mid-morning at our hotel, which was across the street from the Yeltsin "White House." Over 100 thousand people were coming out of subways , busses and walking to the long stairway leading to the Yeltsin headquarters. Yeltsin had declared it a victory day, although it was also a traditional holiday. The Yeltsin coup had been just a week earlier, memorial flowers were still being added to curbside places where soldiers had been killed . Tank barricades were still standing. And the crowds moved festively around them. At the top of the long stairway, singers, and dancers performed. Then speeches from government leaders, echoed over the crowds and down the streets. It was thrilling to hear one of our staff make the official presentation of our Bibles, over the massive loud speakers. Our staff had stashed a big pile of the Bible cartons in a secret place in a nearby hotel lobby. We were told not to let the crowds know where the Bibles were, lest they storm the lobby for them.. So each of us grabbed a carton of 24 Bibles and went to various locations. Before we could open the cartons, people of all ages were tugging at us, even snatching the Bibles in minutes. It was hard to find a quiet place where we could select an individual and nicely present it to them with a " May God bless you". That kind of token distribution was not in the overall plan. But it was awesome to see the people desperately clamoring for a Bible that had been denied them for so many years. We prayed they would read them as fervently. Some of them were reading right on the streets.

The rest of the week we still had appointments to offer portions of the shipment to Christian leaders in Moscow and Kiev, Ukraine And to organizations such as the Russian Orthodox Church, the Moscow Baptist Church, even some TV stations that were reaching people for Christ. The requirement was that the Bibles would be given free. Flying to Kiev on Aeroflot was an experience. While there, our Ukrainian guide, cried as she told us, how as a girl in WWII, she was with her mother, when they escaped from the German train taking them away from Kiev. The guide also took us out to Chernobyl to give presents, candies and Bibles to the young children who suffered radiation from the disaster at the nuclear plant. Heart breaking.

On the way home from Moscow, Gwen and I went to St. Petersburg, to visit the famous Hermitage, and to cruise the River Neva. When we got off the boat, a man came up to us and politely enquired if he could ask a question, since he had overheard us talking about Bibles while on the boat. He stated he was an actor, and had been given an English New Testament. He spoke English fairly well; his question was, " I have been reading that English New Testament; why does it speak to my heart ?" A great point to start at. We answered more questions, as we walked with him to our hotel, where I gave him some helpful literature I had brought.

That night we visited with an inspiring Russian couple, Natalia and Demetri, who had recently been converted to Christ. "You are an answer to prayer" I told them, as I gave them the actor's address to follow him up. At the airport the next day, customs would not let us take any rubles out of the country, but they said if we knew of some one to give it to, they would not keep it, but would let any other Russian have it. It was a considerable amount so I said, "Sure, I have the address of Natalia and Demetri, but how can I get the rubles to them?" Just then a lady behind me spoke up and said, "I can take it to them, if you like, as I live near them." This Russian lady looked sincere and honest so we handed her the rubles. A few weeks later, a letter from Natalia came to us. In it she said, "About the money, thank you. The lady came to give us the rubles, and I talked to her about the Bible and Christ. She stayed two hours and then accepted Jesus as her Savior." Isn't it wonderful how God works?

At our 1991 Board meeting, the mantle of Chairman fell on me. The ministry of the International Bible Society was growing very well, but some of our 19 board members were unhappy with, our president. A key employee was feeding gossip stories about him, secretly to a few board members. The board sent Stan Black and me out to investigate the conflict, and we talked to them both to get to the bottom of it all. Our conclusion was that there was insubordination on the part of the complainer, and though the president could not control that individual, the president was not perpetrating any wrong. But there was a bitter personal conflict. So it was recommended that the complainer be discharged, and when that happened, the fireworks began. Now, we had a board, almost one half of whom wanted to discharge the president as well. After several tense meetings, when the vote was taken, a one-vote majority voted to release the president, much to my displeasure. This was a Saturday afternoon, and the next question was," Who will meet with the 110 dedicated Christian staff on Monday morning and explain the president's departure.? And who will be our next president?"

ENJOYING COLORODO SPRINGS

The board voted that I be elected interim resident – CEO and fly out the following day to Colorado Springs for the Monday meeting with the staff. With a very heavy heart about the situation, I talked candidly to the staff, most of whom loved their discharged president too. But I explained that the International Bible Society was not a business owned by some individuals. It was a "not for profit" organization allowed by the government to operate under the complete control of the Board of Directors, and it was they who could call the shots. I wanted to dispel the notion of some that the " bad board in New York did it" and "what would it mean for their future growth of the NIV". I came out alive from that meeting, and felt the staff would be behind me in the transition. I encouraged dialogue with anyone who wished to enter my office to talk further. Only one person came, but the others did want to know if the board had been fair in the termination arrangements of salary, etc., with the former president. It had been fair. We continued to have Monday morning meetings. It was uplifting to attend them, as there was always prayer, and a devotional message, and it was exciting to see these splendid staffers meeting this way.

We knew that the termination news would rock the so called "Bible" world, as the former president had been so widely known and respected. We also expected repercussions from the media, not only Christian magazines, but also the local daily paper. Therefore I ordered all inquiries on the phone, on the issue, be directed to me, so there would be one voice, and there would be no inadvertent leaks. The reason was to protect the former president's reputation. We therefore let it be known that the termination was not due to wrongdoing or illicit sin, but that there were board issues that had led to the resignation. Our staff responded well to this approach. However, the roof caved in when a full-blown article appeared in the city newspaper. And it did not come from our staff, but I had other suspicions of the source. Needless to say, the former president was further hurt by this, as it made it difficult for him to find a new key job in other

Christian organizations. I kept in touch with him over these matters, and let him keep the company car to facilitate his search for employment. Meanwhile, I moved into the president's office with the big glass window looking across a huge valley to the sprawling site of the Air Force Academy. It was a priceless view because Pikes Peak was also a dominant part of it. But I really never had much time to enjoy it, the pressure of the job was so great. I continued to fly home weekends, as it cost about the same as staying in hotel with meals.

The board suggested that I find an apartment, and live in Colorado Springs until they found a new president. I found a good place and Gwen moved out. By this time the bright young women staff all wanted to see my Gwen as they had been asking questions about her They had a great time one afternoon when Gwen came by the office to meet them, and I could show off my lovely honey to them. It wasn't long before the big city bank called and said eight people were coming out to review our financial standing and future, in the light of the departure of our past president. So one morning my office was crowded with chairs, as both men and women officers of the bank that held our $1,500,000 line of credit, fired questions concerning our current progress and our future. When they finished those questions, the enquiries took on a more spiritual tone. How did the NIV come to be translated? Why was it needed? Why is it becoming so popular among most denominations? All sorts of questions came which literally gave me a platform to also present the Savior of the Bible. I remember quoting John 20: 31, "For these were written, that...". My secretary heard it all, and was thrilled too. And we got our line of credit renewed in a few days, and the Lord was honored.

One day the phone rang "This is Charles Robertson, calling from Johannesburg, South Africa. I would like to come up and talk to you about using the NIV text in a film of the whole Bible." I said "that is very interesting, do you have to fly all the way for that?" He went on to state he wished to review the whole matter with us, and would have his partner from New Zealand come with him. In about two weeks' time they walked into my office. We read together I John 1:1 " That which was from the beginning, which we have heard, which we have seen with our eyes, which our hands have handled, which we have looked at," etc. The two men stated that if we could give them a contract to use the text of the NIV as the script of all the scenes, they intended to produce the whole Bible on film. Each book of the Bible would be filmed and scripted as it is written. It would not be in story form, it would include everything, and only what the text says. This sounded so exciting, it was worth considering approval with a minimum royalty expense to them. We had been cautious because another similar group from Australia had started a similar effort, raised a lot of money from donors, and went bankrupt, leaving some widows holding the bag. We did not want any involvement, with a chance of that happening again. They had $1 million to start up the project, so we granted them the necessary rights to use the NIV. Within the year, they had hired actors of experience to handle the key roles, such as Jesus, the Apostles, etc. They were shooting scenes of Mathew in Morocco, as they could find the semi-desert-like terrain, and the local people as stand ins, looked somewhat like the Palestine countryside natives. When the filming of Mathew was done, they had spent $5 million, and were staggered as to where the money would come from to proceed to another Bible book.

A year later they commenced filming the book of Acts. This time, they chose Tunisia, as the terrain, in which to shoot, as it was similar to Asia Minor. Charles Robertson called me at my home one day from Tunisia, saying "come on down, I'll put you on the set as a native" I told him I saw enough of Tunisia from my flying days there. When they finished Acts, another $5 million had been spent, and they ran out of funds. The initial response to the VCR tapes of Mathew and Acts did not generate sufficient cash to continue, but the quality of the acting and the scene photography has received high marks. The major negative that some people have raised, is that the Jesus of the film is too friendly, smiles too much, and didn't portray the somber Jesus they expected. Now, Mathew and Acts can be purchased on DVDs for use in Bible studies and those who wish to see a good enactment of the " Word of Life" Charles Robertson continued to be a good friend. One occasion when he flew up from down under, he called me at the lake from Nashville. He said he'd like to see me .I warned him of the remoteness of Speculator, New York, but it did not phase him a bit. The next day, he was at our door on the lake. He came by air and rented car. We visited and though I had a room, he would not stay overnight. Unless I missed it, it seemed that he just wanted to visit, although he was rethinking the viability of going further on additional filming.

I received a phone call from the president of the Zondervan Corporation, our principal publisher. He wished me well in my present role at IBS. He was pleased with the way the NIV Bibles were selling to their bookstore customers nationally. But he took the opportunity to jog me on one matter on which our board had been dragging their feet. It was the idea of providing an NIV-like translation at fourth grade reading level, for semi- literates, prisoners, and children. One reason for the great popularity of the NIV was that it was easy to read, at 8^{th}eighth⁻ grade reading level. But children, prisoners, and many under privileged folks, found the NIV sometimes difficult, as also is the King James, at twelfth- grade level.

One of our finest board members, Ron Youngblood, was also on the independent NIV translation group of Bible scholars, still known as "the Committee on Bible Translation" So he was our liaison, to that committee and advisor on such matters. Up to now, we all sensed there was a reluctance to bring out another NIV with a lower vocabulary level, because of the fear of compromising the integrity of the regular NIV. However, Word, and other Bible publishers were coming out with such new approaches. I called in John Cruz, head of our own publishing, discussed the whole matter and concluded that we needed to talk further with Ron Youngblood, the most knowledgeable board member on the issue. And we really needed his blessing, before we could talk to the whole board about the urgency of taking some steps forward with the " readers" Bible concept. When I called Ron I found him to be open about the whole matter, and he agreed that time was now running out, as it would take at least four years to bring out a new children's, or readers version. I stated that I could get quickly our small executive board on a telephone conference call meeting on this subject, to get preliminary approvals for the whole idea, and specifically to approve an initial budget of $600,000, just to get started. He agreed to that and when we did have the conference call meeting, he assured the board it was now the time to get started. Approval was granted by the executive board to move forward now. Fortunately, a very talented scholarly woman, a Children's book writer and PhD, was available in Colorado Springs. She was a close friend of a dedicated staff person, Karen Rayer. She eventually was chosen to "rewrite" the NIV. in fourth grade vocabulary. Then the "Committee on Bible Translation" would again rewrite it, so it would conform to the original manuscripts, and meet the standards of a translation at a reading level of fourth grade. Four years later, when the NIV "Readers" Bible got to the book stores, prisons, etc., I was happy to learn it was being received well throughout the country. Thank you, Lord. Ephesians 2:10.

A letter came to my office from Hodder and Stoughton in London. It was an invitation for Gwen and me to join with them in May to present the new Archbishop of Canterbury with an NIV Bible. The Rt. Rev. Archbishop George Carey had agreed to a ceremonial presentation at Lambeth Palace, across the Thames river from Parliament. The board gave me a hearty approval to make the trip, and offered to pay all our expenses, as a thank you, since I had not taken any remuneration for my away from home CEO duties at IBS while in Colorado. Gwen had to buy new clothes for the occasion, and she even bought a cute red straw hat with a turned-up brim. We stayed in a lovely hotel, but were forced out of our rooms twice by sirens, and emergency departures to the lobby. Both cases were due to suspicious items, one a bag, and the other a parked car on the curb in front of the adjacent building. The terrifying Irish Republican Army at work. In the morning, we were fortunate to see the Queen's entourage proceeding along the streets on the opening day of Parliament. Crowds lined the streets to see her pass by in the royal carriage. We took a lot of pictures, but none of them could be developed due to a camera defect.

Our appointment at Lambeth Palace was at noon. Gwen looked great in her cute red hat and pretty new dress. We were picked up in a limousine by the top officers of Hodder and

Stoughton, and we drove across the Thames to the long wide steps leading to the huge guarded doors of Lambeth Palace. It was a palatial looking building that for hundreds of years has been the gathering place of Episcopal and Anglican clergy from all over the world. I had read several small books about George Carey, his early life and his testimony of being convicted of his need to be born again before he could really be of service to the Lord.

So it was a thrill to walk up to the fireplace where he was standing with his wife, and shake their hands. Gwen noticed that Mrs.Carey did not wear a hat, so she discarded the pretty red hat quickly. We did not expect a luncheon, but they did serve some wine, crackers and cheese, which gave us some pleasant time to visit. When it was time to make the presentation of the NIV, the Bible box was opened. It was the most beautiful deep red leather cover; it had been specially made in Scotland . The spine letters and the page edges were lit up with shiny gold. The red leather box was lined with velvet. It looked like a ceremonial Bible for the Archbishop's library. In my presentation, I expressed how pleased we were in America and in England, that he was recognizing the NIV in this way. Of course, we were not seeking his endorsement and it would not alter the long traditional use of the King James' Version in Britain's churches, but it would add to the credibility of the NIV in bookstores, and the public, plus ministries like the British Gideon's, when this presentation got out in the press. Actually, the British Gideons were already using the NIV, whereas the American Gideons were not.

Before we left, George Carey went back to his office, came back to give me two autographed copies of his books. And as our party walked down the wide steps, I showed them the signature that the Archbishop placed in the books. " It is not written as George Carey" I said "he signed it as George Canter." "Oh yes," our hosts said" It is a long tradition that the Archbishop of Canterbury changes his last name to Canter, while he is in office"

On the trip home, Gwen proceeded through customs with her bags, with her pretty hat on. All our bags got through inspection, except Gwen's hat, which when the agent saw it on her, asked for it to be removed for inspection. Then they proceeded to go over that hat as though there was a bomb in it. We laughed all the way home.

An invitation came from Jim Dobson one day shortly after Focus on the Family had moved from California to Colorado Springs. They brought 600 people and would be hiring more. They had found office space for the staff while their new quarters was being built, and it meant that they had to rent several office buildings in the downtown city. Jim Dobson wanted to have the Presidents of the major Christian organizations for a luncheon at his office downtown. Fifteen attended and it was a good office catered luncheon. Jim Dobson spoke about their new building plans, and Shirley spoke about the soon to be National Prayer Day, of which she had just been appointed chairperson. Then Jim asked each ministry leader to introduce themselves, their organization name, and the mission statement that each uses to describe their ministry purpose. Some could quote their actual mission statement, others gave a summary in a few sentences; when it came my turn , I was fortunate to have a brand new calling card, and on the backside of it was our new mission statement, which our board had just approved! After Focus completed the new headquarters, they doubled the size of their staff and put millions into their beautiful site. Tours and all.

We made Northeast Bible Chapel our temporary home church, while in Colorado Springs. There were three Brethren assemblies there, all growing rapidly. Some wonderful Christians there, too. One was a woman who also worked at IBS, a Japanese widow, who was a fantastic person. Her "gift" was hospitality, not only to church folks, but mainly to Air Force cadets of Asian decent. If they were Christians or not, she drove out to the Academy on Sunday mornings, to take them to the services at the church. Then she cooked dinner for them and let her home be their home for the day. She has been doing it so many years, that graduates who left, are now inviting her to their weddings, or to meet their families, as though she was still their "mother".

One Sunday after the Lord's Supper, a person came up to me and said "I saw your name in the newspaper this morning," In checking it out, I found a full page invitation to the dedication of the huge New Life Fellowship building that was on the property across from our IBS headquarters. Noted speaker, Jack Hayford, author of the popular chorus "Majesty," was coming from California to bring the dedication message. Also on the page, there was a list of 10 local guests, who included James Dobson, and a few others I knew. When we arrived, we were ushered to special front section of the enormous but plain, painted cement block sanctuary. Ted Haggard , the pastor, was then a rising star, in the evangelical "world" (sad to say, this was the Haggard whose Christian service tragically ended in 2006, due to gross sin) At that time, his super extroverted personality was getting crowds of several thousand members. Jack Hayford gave an excellent message on "The Glory of God in the Home". The next day I ran into James Dobson's #2 man, and asked him how Jim liked the service, to which he smiled and said "he wondered a few times how he got into that semi-charismatic situation, but it was okay."

There was some time for diversion and fun. A fellow from my home town, who was visiting his son, an instructing officer at the Air Force Academy, invited me to an Air Force basket-ball game followed by an ice hockey game, a double header against Duke. We bought programs, and at the hockey game, surprise, the winning number on the program cover was my number. This meant I could go down on the ice, and with a hockey stick and a puck, make three shots at a box in the net. If the puck went in, the prize was $5,000. When I got to the ice the emcee asked a lot of questions: like what I do, where I live, etc. I also mentioned that I was an ex-Air Corp pilot of WWII, etc. When I stepped on the rink to shoot, I got a tremendous applause, which I could not figure out. But I shot the three pucks and lost, returned to my seat, and my friend said," That was great about the P-40s and the P-47s." I said "what"? And it then dawned on me that the emcee had his mike on when I thought we were having a private conversation. But I was thrilled, in retrospect, to get an ovation from Air Force cadets, whom I have always admired.

There were great times of fun and fellowship with our board members, too. Several times, arrangements were made to spend several days at a fantastic Dude ranch, in central Colorado, called

Lost Valley Ranch. It was a pretty setting 9 miles in from a highway, at 6,500 feet altitude, in John Wayne-like country. They had over 100 riding horses for guests, and lovely cabins with a fireplace. It was owned by the Fosters, great Christian folks who loved the Lord and showed it. Their clientele came from all over the country. Even Walt Disney put his ranch branding sign on the lobby wall. They did not run it as a so-called Christian ranch, but all his employees were young people from Christian colleges, Navigators, Youth for Christ or Inter Varsity Christian Fellowship. They were his badge of honor, so when non-Christian guests would ask Foster where he got such fine employees to wait on tables and groom the horses, he would tell them of their background and faith, and open doors for other conversations about the Lord.. It was great to ride those beautiful stallions up hillsides and down in gullies, places that the horses seemed to always have control. Most of us old timers rode horses for the first time, and got a thrill every time. But the others just wore the big ranch hats, and went walking up the path along the rushing cold stream that watered the horses, and winded through the attractive valley of the ranch. Gwen and other ladies hiked the paths along the stream to where the horses drank.

Meanwhile, candidates for a new president were being interviewed by the search committee of the board, some of whom lived in Colorado. No good leads. However, at the next board meeting, it was suggested we consider following up contacts that had been made with the Director of the international wing of the Living Bible organization. Several months prior, they expressed an interest in some form of working together on large foreign translations. IBS had only the NIV in English, but were contemplating an expansion into many places countries, greater than one million, where Bibles in the native language was not available. So when we approached them, we found a genuine willingness to explore the options. And since Living Bibles International (LBI) had a fine President, Lars Dunberg, with great experience in the Bible world, we soon began exploring the feasibility of a merger between LBI and IBS. If this was possible, it would solve our search for a good leader.

When a committee of both parties met, we were impressed with LBI's board and their commitment to publishing scriptures. On their board was Ken Taylor, author of the popular paraphrased Living Bible (the Living Bible was not part of the merger) Mark Taylor, President of Tyndale Publishers, Lars, Luci Shaw, Louis Palau, evangelist, and others. We soon began talking about a full merger, in which no money would be involved, in which all their assets and liabilities would be blended with ours, and all their board members would be taken on our board, including Lars. This would fit well too, as our board had a number of older men who would be replaced in the near future, although we had some splendid younger men like Dave Sunden, John Faust and Stan Black who could carry on well. The LBI people were even willing to give up the LBI name and go on the name of IBS.

Both boards eventually approved the idea, and we proceeded to run "due diligence" investigations of their assets (up to $20 million in translations) and liabilities (workers and small corporate entities in countries where Bibles were sold) and over a million mortgage on their Wheaton building.

Fortunately, we had Stan Black, an astute lawyer from Boulder, on our board He guided us through many decisions and legal hurdles to execute a contract. Since ours was a New York corporation, and theirs was a Wheaton, Illinois corporation, and we were forming another tax-free charitable corporation in Colorado, you can imagine the paper work, approvals and even confusion this would make. I was 73 then, retired from my own business and I was working harder than any time in my life, but enjoying it every new day. Stan Black's legal work made the difference in unlocking many of the hurdles, right there in Colorado. When I resigned at the required level of 75, the board wisely voted Stan Black as our new Chairman. Good choice, good man.

50th ANNIVERSARY   50th REUNION

It was somewhat of a letdown to return to "retired life" again. Our church-related interests kept us busy, though with less responsibilities, except as an elder.   In October, 1994 Gwen looked her loveliest, as we celebrated our 50th wedding anniversary.   Barbara, Patsy, John and Betty arranged a great program at the Molly Pitcher Hotel, overlooking the beautiful Navesink River.   Contributing to the program too, was Barbara's husband, Jack Gill, Patsy's husband Ted Drab, and John's wife, Joyce.   Plus all of our lovely grandchildren, and relatives from the Dunkerton and the Pinkham side and some very close friends, too.   Pictures now on DVD tell the story of one of the happiest days of our lives.

Another memorable occasion was a trip to the Oshkosh, Wisconsin, National Air Show, put on by the small aircraft industry, known as AOPA.   Some 20,000 take-offs and landings are made during the week or so the small aircraft came and went to the show.   Every day acrobatics were performed directly over the field.   Between those events, a performance fly over of military B-2 bombers, P-51s or Harriots, the vertical take-off jet fighter, captured the attention of the crowds.   It seemed every minute was occupied with an air event.   One day, the loud speaker announced a demonstration of the short take-off and landing techniques of a plane used by the Wycliffe Translators. The pilot landed and rolled only the length of the airplane, and took off in the same space.   The announcer gave Wycliffe a good testimony too, as to what they do and where on the field to learn more about them.

However, the most exciting thing happened when I purchased a colored book of P-40 planes and the pilots who flew them.   I noticed that four pages were of men from my squadron, all of whom I had known since they were with the squadron at the same time we were in combat.   When I got home I wrote the author, Jeff Ethell, asking for addresses of my guys.   After he called to give me their phone numbers, I spent an exciting morning talking to men I had lost contact with for 50 years. The most thrilling was to know that Jim Pittard was alive, as it was thought he had died in the German prison camp. And Jaslow was fine, had retired as a dentist. Duke was dead. They told me about reunions which had brought out all the enlisted men, as well as the pilots of the 79th Fighter Group together. So I mailed Jim Pittard's whittling knife to him and apologized for not sending it to his wife Betty, in 1943 when he was shot down. And I went to reunions several years in a row. But many of the pilots of the early struggle in the Mediterranean theater, stopped coming out. So I never saw most of the pilots with whom I had flown missions. But there were about 100 of the former mechanics, radio, meteorology, intelligence, armament and other service fields that kept a squadron going. It was great to see Goose Gossick again. Back then, he was Goose, though he was a captain. But now he was General Gossick, and I did not dare to call him Goose. He remained in the Air Force in engineering development, was the commanding general in charge of the Tullahoma Space Engine testing base. He used his clout to allow our reunion to be held there, so we were treated to all the wonders of aviation testing. The most remarkable being the 3 million gallon per minute, water cooling chambers for space engines, and the wind tunnel evaluator, which both military and Boeing used to test their 3-feet-long precisely scaled models. Jesse Jory also attended, and we had a lot of fun recalling the flights of the German ME-109 he had put together for us. So did Colonel Bates, our former 79th Fighter Group Commander, who still commanded the esteem we always had for him.

A while later, I told Gordon Purdy the above incident involving Jeff Ethell, author of the P-40 book. Gordon, a Cessna-172 owner, and a real aviation enthusiast himself, then told me of his friendship with Ethell. In fact, Gordon had invited Ethell to be the featured guest speaker at the previous spring" Fly In" into Camp of the Woods. The annual "Fly In" brings in scores of pontoon planes from all over the northeast. They land on Lake Pleasant, and taxi in to the quarter-mile shallow sand beach, for the weekend. Many retired plane owners drove in their cars, and filled all the motels in the area. Camp of the Woods accommodated all the pilots and guests on condition that the "Fly In" group have a Sunday service. Some years that service was done by Moody Aviation People, but this year Gordon had invited Jeff Ethell, not only because of his aviation books and accomplishments, but mainly because of his genuine Christian life for his Lord. An outspoken born again spokesman for Christ. However, Gordon stated, " Two weeks before the fly In, Jeff Ethell was flying a P-38 WW2 fighter plane, when something happened, and he augured in, out of control, to his death. Yet, God is sovereign, and Jeff is with his Lord.

Though I have flown through the experiences of life on cloud nine, and also walked through the valley of death, I can honestly say that "the Lord is my Light and my salvation, whom then shall I fear, of whom shall I be afraid." (Psalm 27:1) I do believe in God's great love. And I also believe in God's presence in one's life. That occurs when one places his trust in God's way and God's Son, for the salvation of the soul. Step one is to realize the basic need of a Savior, because "all have sinned" and "we have turned to our own way". But the good news is that God loved us so much, He "laid on Him, the iniquity of us all" These are quotes from scripture, but this is how God speaks. Through his Son, by His word. God sent Christ into our world to reveal his plan for saving us on the cross.

And perhaps it is odd to us that he didn't say do your best, or follow certain rules. But Christ said "I am the resurrection and the life. He that believeth in me, shall never die, but shall have everlasting life." So when Christ died at the hands of sinful men, it was not God's plan to let him be just a martyr, and let the world just admire a selfless life, and simply morn his death endlessly. Not at all, Christ came to reveal his Father's love for us, and have His Son "bear our sins in his own body" and thus be satisfied that the most perfect sacrifice was made for everyone. So when you or I come to accept God's Son, who is God's remedy for cleansing our lives from sin, then God is pleased with you, and thus He will save you now, and bring you to His presence when He calls you to His home. Isn't that wonderful.

So does God want His people to do good works? You bet He does. Read Ephesians 2:8-10, for "it is by grace you have been saved, through faith, and this is not of yourselves, it is the gift of God---not by works, so that no one can boast. We are God's workmanship, created in Christ Jesus to do good works, which God prepared in advanced for us to do." So not only can we please Him in our lives as we live for Him, but we will praise Him in   heaven with millions of others. For we have a guaranteed hope, and a sure promise of eternal life with our Savior. Read Colossians 1: 9-15, as this is my prayer for you.

Jack and Gwen at Willow Valley on their 63rd wedding anniversary.